50% OFF Online SAT Prep Course!

Dear Customer,

We consider it an honor and a privilege that you chose our SAT Study Guide. As a way of showing our appreciation and to help us better serve you, we have partnered with Mometrix Test Preparation to offer you **50% off their online SAT Prep Course.** Many SAT courses are needlessly expensive and don't deliver enough value. With this course, you get access to the best SAT prep material, and you only pay half price.

Mometrix has structured their online course to perfectly complement your printed study guide. The SAT Prep Course contains **in-depth lessons** that cover all the most important topics, **120+ video reviews** that explain difficult concepts, over **800 practice questions** to ensure you feel prepared, and more than **400 digital flashcards**, so you can study while you're on the go.

Online SAT Prep Course

Topics Covered:

- Reading
  - Literary Analysis
  - Inferences in a Text or Texts
  - Author's Use of Language
- Writing
  - Foundations of Grammar
  - Essay Revision and Sentence Logic
  - Agreement and Sentence Structure
- Mathematics
  - Algebra and Functions
  - Geometry and Measurement
  - Data Analysis, Statistics, and Probability

Course Features:

- SAT Study Guide
  - Get content that complements our best-selling study guide.
- Full-Length Practice Tests
  - With over 800 practice questions, you can test yourself again and again.
- Mobile Friendly
  - If you need to study on the go, the course is easily accessible from your mobile device.
- SAT Flashcards
  - Our course includes a flashcards mode with over 400 content cards for you to study.

To receive this discount, visit their website: mometrix.com/university/sat and add the course to your cart. At the checkout page, enter the discount code: **APEXSAT50**

If you have any questions or concerns, please contact them at universityhelp@mometrix.com.

Sincerely,

APEX test prep in partnership with Mometrix TEST PREPARATION

FREE

Free Study Tips DVD

In addition to the tips and content in this guide, we have created a FREE DVD with helpful study tips to further assist your exam preparation. **This FREE Study Tips DVD provides you with top-notch tips to conquer your exam and reach your goals.**

Our simple request in exchange for the strategy-packed DVD is that you email us your feedback about our study guide. We would love to hear what you thought about the guide, and we welcome any and all feedback—positive, negative, or neutral. It is our #1 goal to provide you with top quality products and customer service.

To receive your **FREE Study Tips DVD**, email freedvd@apexprep.com. Please put "FREE DVD" in the subject line and put the following in the email:

a. The name of the study guide you purchased.

b. Your rating of the study guide on a scale of 1-5, with 5 being the highest score.

c. Any thoughts or feedback about your study guide.

d. Your first and last name and your mailing address, so we know where to send your free DVD!

Thank you!

SAT Prep 2021 and 2022 with Practice Tests

SAT Study Guide for the College Board Exam [4th Edition Book]

Matthew Lanni

Written and edited by APEX Publishing.

ISBN 13: 9781628457858
ISBN 10: 1628457856

For additional information or for bulk orders, contact info@apexprep.com.

Table of Contents

Test Taking Strategies

1. Reading the Whole Question

A popular assumption in Western culture is the idea that we don't have enough time for anything. We speed while driving to work, we want to read an assignment for class as quickly as possible, or we want the line in the supermarket to dwindle faster. However, speeding through such events robs us from being able to thoroughly appreciate and understand what's happening around us. While taking a timed test, the feeling one might have while reading a question is to find the correct answer as quickly as possible. Although pace is important, don't let it deter you from reading the whole question. Test writers know how to subtly change a test question toward the end in various ways, such as adding a negative or changing focus. If the question has a passage, carefully read the whole passage as well before moving on to the questions. This will help you process the information in the passage rather than worrying about the questions you've just read and where to find them. A thorough understanding of the passage or question is an important way for test takers to be able to succeed on an exam.

2. Examining Every Answer Choice

Let's say we're at the market buying apples. The first apple we see on top of the heap may *look* like the best apple, but if we turn it over we can see bruising on the skin. We must examine several apples before deciding which apple is the best. Finding the correct answer choice is like finding the best apple. Although it's tempting to choose an answer that seems correct at first without reading the others, it's important to read each answer choice thoroughly before making a final decision on the answer. The aim of a test writer might be to get as close as possible to the correct answer, so watch out for subtle words that may indicate an answer is incorrect. Once the correct answer choice is selected, read the question again and the answer in response to make sure all your bases are covered.

3. Eliminating Wrong Answer Choices

Sometimes we become paralyzed when we are confronted with too many choices. Which frozen yogurt flavor is the tastiest? Which pair of shoes look the best with this outfit? What type of car will fill my needs as a consumer? If you are unsure of which answer would be the best to choose, it may help to use process of elimination. We use "filtering" all the time on sites such as eBay® or Craigslist® to eliminate the ads that are not right for us. We can do the same thing on an exam. Process of elimination is crossing out the answer choices we know for sure are wrong and leaving the ones that might be correct. It may help to cover up the incorrect answer choice. Covering incorrect choices is a psychological act that alleviates stress due to the brain being exposed to a smaller amount of information. Choosing between two answer choices is much easier than choosing between all of them, and you have a better chance of selecting the correct answer if you have less to focus on.

4. Sticking to the World of the Question

When we are attempting to answer questions, our minds will often wander away from the question and what it is asking. We begin to see answer choices that are true in the real world instead of true in the world of the question. It may be helpful to think of each test question as its own little world. This world may be different from ours. This world may know as a truth that the chicken came before the egg or may assert that two plus two equals five. Remember that, no matter what hypothetical nonsense may be in the question, assume it to be true. If the question states that the chicken came before the egg, then choose your answer based on that truth. Sticking to the world of the question means placing all of our biases and assumptions aside and relying on the question to guide us to the correct answer. If we are simply looking for answers that are correct based on our own judgment, then we may choose incorrectly. Remember an answer that is true does not necessarily answer the question.

5. Key Words

If you come across a complex test question that you have to read over and over again, try pulling out some key words from the question in order to understand what exactly it is asking. Key words may be words that surround the question, such as *main idea, analogous, parallel, resembles, structured,* or *defines.* The question may be asking for the main idea, or it may be asking you to define something. Deconstructing the sentence may also be helpful in making the question simpler before trying to answer it. This means taking the sentence apart and obtaining meaning in pieces, or separating the question from the foundation of the question. For example, let's look at this question:

> Given the author's description of the content of paleontology in the first paragraph, which of the following is most parallel to what it taught?

The question asks which one of the answers most *parallels* the following information: The *description* of paleontology in the first paragraph. The first step would be to see *how* paleontology is described in the first paragraph. Then, we would find an answer choice that parallels that description. The question seems complex at first, but after we deconstruct it, the answer becomes much more attainable.

6. Subtle Negatives

Negative words in question stems will be words such as *not, but, neither,* or *except.* Test writers often use these words in order to trick unsuspecting test takers into selecting the wrong answer—or, at least, to test their reading comprehension of the question. Many exams will feature the negative words in all caps (*which of the following is NOT an example*), but some questions will add the negative word seamlessly into the sentence. The following is an example of a subtle negative used in a question stem:

> According to the passage, which of the following is *not* considered to be an example of paleontology?

If we rush through the exam, we might skip that tiny word, *not,* inside the question, and choose an answer that is opposite of the correct choice. Again, it's important to read the question fully, and double check for any words that may negate the statement in any way.

7. Spotting the Hedges

The word "hedging" refers to language that remains vague or avoids absolute terminology. Absolute terminology consists of words like *always, never, all, every, just, only, none,* and *must.* Hedging refers to words like *seem, tend, might, most, some, sometimes, perhaps, possibly, probability,* and *often.* In some cases, we want to choose answer choices that use hedging and avoid answer choices that use absolute terminology. It's important to pay attention to what subject you are on and adjust your response accordingly.

8. Restating to Understand

Every now and then we come across questions that we don't understand. The language may be too complex, or the question is structured in a way that is meant to confuse the test taker. When you come across a question like this, it may be worth your time to rewrite or restate the question in your own words in order to understand it better. For example, let's look at the following complicated question:

> Which of the following words, if substituted for the word *parochial* in the first paragraph, would LEAST change the meaning of the sentence?

Let's restate the question in order to understand it better. We know that they want the word *parochial* replaced. We also know that this new word would "least" or "not" change the meaning of the sentence. Now let's try the sentence again:

> Which word could we replace with *parochial,* and it would not change the meaning?

Restating it this way, we see that the question is asking for a synonym. Now, let's restate the question so we can answer it better:

> Which word is a synonym for the word *parochial*?

Before we even look at the answer choices, we have a simpler, restated version of a complicated question.

9. Predicting the Answer

After you read the question, try predicting the answer *before* reading the answer choices. By formulating an answer in your mind, you will be less likely to be distracted by any wrong answer choices. Using predictions will also help you feel more confident in the answer choice you select. Once you've chosen your answer, go back and reread the question and answer choices to make sure you have the best fit. If you have no idea what the answer may be for a particular question, forego using this strategy.

10. Avoiding Patterns

One popular myth in grade school relating to standardized testing is that test writers will often put multiple-choice answers in patterns. A runoff example of this kind of thinking is that the most common answer choice is "C," with "B" following close behind. Or, some will advocate certain made-up word patterns that simply do not exist. Test writers do not arrange their correct answer choices in any kind of pattern; their choices are randomized. There may even be times where the correct answer choice will be the same letter for two or three questions in a row, but we have no way of knowing when or if this might happen. Instead of trying to figure out what choice the test writer probably set as being correct, focus on what the *best answer choice* would be out of the answers you are presented with. Use the tips above, general knowledge, and reading comprehension skills in order to best answer the question, rather than looking for patterns that do not exist.

FREE DVD OFFER

Achieving a high score on your exam depends not only on understanding the content, but also on understanding how to apply your knowledge and your command of test taking strategies. **Because your success is our primary goal, we offer a FREE Study Tips DVD, which provides top-notch test taking strategies to help you optimize your testing experience.**

Our simple request in exchange for the strategy-packed DVD is that you email us your feedback about our study guide.

To receive your **FREE Study Tips DVD**, email freedvd@apexprep.com. Please put "FREE DVD" in the subject line and put the following in the email:

a. The name of the study guide you purchased.

b. Your rating of the study guide on a scale of 1-5, with 5 being the highest score.

c. Any thoughts or feedback about your study guide.

d. Your first and last name and your mailing address, so we know where to send your free DVD!

Introduction to the SAT

Function of the Test

The SAT is an exam used by a range of universities to determine entrance into college. It is an exam given by the College Board, a non-profit organization that exists to serve higher education. College admissions use the test score to assess a high school student's college readiness along with their high school GPA, extracurricular activities, writing samples, or any other materials used to gain entrance into a university. For the class of 2017, 1.7 million students took the new SAT at least once during their high school career. The mean total score was 1060, on a score ranging from 400 to 1600.

Test Administration

The SAT exam is administered by the Educational Testing Service via the College Board website. On this site, you can register for the exam at a testing center near you. The testing deadlines for 2020 are in July, September, October, November, February, April, and May. The exam is offered seven times in the 2020/21 year. There is no limit to how many times a student can take the SAT. Keep in mind that some schools use a "send all scores" report, so you might want to limit yourself to taking the SAT between two and five times and do your best on those exams to show that you are taking the test seriously. Accommodations regarding disabilities can be requested from the college board website.

Test Format

Test centers for the SAT exam open at 7:45 a.m. and close at 8:00 a.m., so make sure you arrive accordingly. The test starts between 8:00 and 9:00 a.m., where a coordinator will read the instructions and let you know when to begin and end each section. One ten-minute and one five-minute break are allowed during the test. Testing is usually done around 1:00 p.m., if you take the essay. Make sure you bring your admission ticket, an acceptable photo ID, two number 2 pencils, and an approved calculator. Check on the College Board's website to see what kind of materials not to bring.

The SAT is made up of four tests: the Reading Test, the Writing and Language Test, the Math Test, and the SAT Essay. The essay is optional, but some colleges do require you to take it. Note that there is no penalty for guessing. You earn points for the questions you answer correctly, so make sure not to leave any blank. The table below depicts each section along with its number of questions.

Subject	Questions	Time
Reading	52 questions	65 minutes
Writing and Language	44 questions	35 minutes
Math, Calculator	38 questions	55 minutes
Math, No Calculator	20 questions	25 minutes
Essay	1 prompt	50 minutes

Scoring

When you register for the SAT and up to nine days after, you can send four free score reports to the school of your choosing. Scores range from a 400 to 1600. Although there's no such thing as a passing score, there are resources that will let you know what universities expect from your SAT score.

Recent/Future Developments

In June 2021, the optional Essay will no longer be available in many schools, and will no longer be available for students who take the SAT on a Saturday. Some schools may choose to give the optional SAT as part of SAT School Day administrations, but students must check with their school beforehand.

Study Prep Plan for the SAT

1 Breathe
Reducing stress is key when preparing for your test.

2 Build
Create a study plan to help you stay on track.

3 Begin
Stick with your study plan. You've got this!

1 Week Study Plan

Day 1	Day 2	Day 3	Day 4	Day 5	Day 6	Day 7
Reading	Writing and Language	Math	Essay	Practice Test #1	Pracrice Test #2	Take your Exam!

2 Week Study Plan

Day 1	Day 2	Day 3	Day 4	Day 5	Day 6	Day 7
Command of Evidence	Word in Context	Practice Questions	Standard English Conventions	Expression of Ideas	Practice Questions	Heart of Algebra

Day 8	Day 9	Day 10	Day 11	Day 12	Day 13	Day 14
Problem Solving and Data Analysis	Passport to Advanced Math	Additional Topics in Math	Practice Questions	Practice Test #1	Practice Test #2	Take your Exam!

30 Day Study Plan

Day 1	Day 2	Day 3	Day 4	Day 5	Day 6	Day 7
Summarizing a Complex Text	Identifying Specific Information from a Printed Communication	Identifying Information from a Graphic Representation of Information	Recognizing Events in a Sequence	Context Clues	Interpreting the Meaning of Words and Phrases Using Context	Evaluating the Author's Point of View in a Given Text

Day 8	Day 9	Day 10	Day 11	Day 12	Day 13	Day 14
Examining Hypotheses	Practice Questions	Conventions of Standard English Spelling	Elements of the Writing Process	Practice Questions	Translating Phrases and Sentences into Expressions, Equations, and Inequalities	Creating, Solving, or Interpreting Systems of Two Linear Equations in Two Variables

Day 15	Day 16	Day 17	Day 18	Day 19	Day 20	Day 21
Understanding Connections Between Algebraic and Graphical Representations	Using Ratios, Rates, Proportional Relationships, and Scale Drawings...	Using the Relationship Between Two Variables to Investigate Key Feature...	Using Two-Way Tables to Summarize Categorical Data and Relative Frequencies...	Determining the Most Suitable Form of an Expression or Equation	Solving a System of One Linear Equation and One Quadratic Equation	Understanding the Relationship Between Zeros and Factors of Polynomials

Day 22	Day 23	Day 24	Day 25	Day 26	Day 27	Day 28
Solving Problems Using Volume Formulas	Converting Between Degrees and Radians, and Using Radians to Determine Arc Lengths	Using Concepts and Theorems about Congruence and Similarity...	Practice Questions	Practice Test #1	Review Answer Explanations	Practice Test #2

Day 29	Day 30
Review Answer Explanations	Take your Exam!

SAT Reading Test

Command of Evidence

Summarizing a Complex Text

An important skill is the ability to read a complex text and then reduce its length and complexity by focusing on the key events and details. A **summary** is a shortened version of the original text, written by the reader in their own words. The summary should be shorter than the original text, and it must include the most critical points.

In order to effectively summarize a complex text, it's necessary to understand the original source and identify the major points covered. It may be helpful to outline the original text to get the big picture and avoid getting bogged down in the minor details. For example, a summary wouldn't include a statistic from the original source unless it was the major focus of the text. It's also important for readers to use their own words yet retain the original meaning of the passage. The key to a good summary is emphasizing the main idea without changing the focus of the original information.

Complex texts will likely be more difficult to summarize. Readers must evaluate all points from the original source, filter out the unnecessary details, and maintain only the essential ideas. The summary often mirrors the original text's organizational structure. For example, in a problem-solution text structure, the author typically presents readers with a problem and then develops solutions through the course of the text. An effective summary would likely retain this general structure, rephrasing the problem and then reporting the most useful or plausible solutions.

Paraphrasing is somewhat similar to summarizing. It calls for the reader to take a small part of the passage and list or describe its main points. Paraphrasing is more than rewording the original passage, though. As with summary, a paraphrase should be written in the reader's own words, while still retaining the meaning of the original source. The main difference between summarizing and paraphrasing is that a summary would be appropriate for a much larger text, while paraphrase might focus on just a few lines of text. Effective paraphrasing will indicate an understanding of the original source, yet still help the reader expand on their interpretation. A paraphrase should neither add new information nor remove essential facts that change the meaning of the source.

Inferring the Logical Conclusion from a Reading Selection

Making an **inference** from a selection means to make an educated guess from the passage read. Inferences should be conclusions based off of sound evidence and reasoning. When multiple-choice test questions ask about the logical conclusion that can be drawn from reading text, the test-taker must identify which choice will unavoidably lead to that conclusion. In order to eliminate the incorrect choices, the test-taker should come up with a hypothetical situation wherein an answer choice is true, but the conclusion is not true.

For example, here is an example with three answer choices:

> Fred purchased the newest PC available on the market. Therefore, he purchased the most expensive PC in the computer store.
>
> What can one assume for this conclusion to follow logically?
>
> a. Fred enjoys purchasing expensive items.
> b. PCs are some of the most expensive personal technology products available.
> c. The newest PC is the most expensive one.

The premise of the text is the first sentence: Fred purchased the newest PC. The conclusion is the second sentence: Fred purchased the most expensive PC. Recent release and price are two different factors; the difference between them is the logical gap. To eliminate the gap, one must equate whatever new information the conclusion introduces with the pertinent information the premise has stated. This example simplifies the process by having only one of each: one must equate product recency with product price. Therefore, a possible bridge to the logical gap could be a sentence stating that the newest PCs always cost the most.

Identifying the Topic, Main Idea, and Supporting Details

The **topic** of a text is the general subject matter. Text topics can usually be expressed in one word, or a few words at most. Additionally, readers should ask themselves what point the author is trying to make. This point is the **main idea** of the text, the one thing the author wants readers to know concerning the topic. Once the author has established the main idea, they will support the main idea by supporting details. **Supporting details** are evidence that support the main idea and include personal testimonies, examples, or statistics.

One analogy for these components and their relationships is that a text is like a well-designed house. The topic is the roof, covering all rooms. The main idea is the frame. The supporting details are the various rooms. To identify the topic of a text, readers can ask themselves what or who the author is writing about in the paragraph. To locate the main idea, readers can ask themselves what one idea the author wants readers to know about the topic. To identify supporting details, readers can put the main idea into question form and ask "what does the author use to prove or explain their main idea?"

Let's look at an example. An author is writing an essay about the Amazon rainforest and trying to convince the audience that more funding should go into protecting the area from deforestation. The author makes the argument stronger by including evidence of the benefits of the rainforest: it provides habitats to a variety of species, it provides much of the earth's oxygen which in turn cleans the atmosphere, and it is the home to medicinal plants that may be the answer to some of the world's deadliest diseases. Here is an outline of the essay looking at topic, main idea, and supporting details:

- Topic: Amazon rainforest
- Main Idea: The Amazon rainforest should receive more funding in order to protect it from deforestation.
- Supporting Details:
    - 1. It provides habitats to a variety of species
    - 2. It provides much of the earth's oxygen which in turn cleans the atmosphere
    - 3. It is home to medicinal plants that may be the answer to some of the world's deadliest diseases.

Notice that the topic of the essay is listed in a few key words: "Amazon rainforest." The main idea tells us what about the topic is important: that the topic should be funded in order to prevent deforestation. Finally, the supporting details are what author relies on to convince the audience to act or to believe in the truth of the main idea.

Following a Given Set of Directions

To follow directions, an individual must hold them in short-term memory while using working memory and language to comprehend the information. Certain word types are found more often than others in directions; these include adjectives, negatives, conjunctions, and prepositions. Understanding the concepts that these word types represent is important, especially in academics. For example, the CELF (Clinical Evaluation of Language Fundamentals) subtest on Concepts and Following Directions is one of most frequently administered assessment instruments for evaluating how well people follow directions, and almost every item on this subtest combines negatives, conjunctions, and prepositions. Being able to follow a direction requires understanding of the direction's specific words. Even single-step directions can be challenging, especially when they contain negatives ("Do *not* circle the even-numbered items"). In addition to language comprehension, the ability to follow directions requires interest in the activity, motivation to follow the directions, and paying attention to the directions.

Basic one-step directions use simple syntax and kid's earliest learned words, like the following:

- Give me the cup
- Tie your shoe
- Say your name
- Close the door
- Throw that away

Expanded one-step directions add a few negatives, contractions, and slightly more advanced vocabulary, like the following:

- Don't touch the blue one
- Show me which one is not yellow
- Use the bigger pen
- Throw all of your trash away
- Take out your small, green notebook

Basic two-step directions include one basic conjunction, or one added phrase, like the following:

- Take off your socks and put them in the hamper
- Close the door and then sit down
- Write your name on the test before beginning
- Give me the glass that is empty

Expanded two-step directions have a structure added to two steps, like the following:

Throw your partner's trash away before you start the assignment

Complex directions include vocabulary and syntax that develop later, plus multiple clauses, like the following:

Please turn off all cell phones after you have taken your seats

The following example shows why it's extremely important to read all the directions in a test setting first before acting on any of them. Follow the directions below:

1. Carefully read everything in the directions before you do anything.
2. Write your name on the top of this page.
3. Put your birthday and date underneath your name.
4. Write an "Z" on an empty space anywhere on the page.
5. Circle the "Z" and draw a line below it referring to "Y."
6. Draw a line from "Y" referring to "X."
7. Draw a box at the bottom of this page.
8. Describe the shape you drew around the letters inside the box.
9. Write your name inside the box.
10. Now that you've carefully read everything, follow only directions (1) and (2).

Those who get to direction 10 without having acted on any instructions will save a lot of wasted time opposed to those who immediately begin to act on the instructions. Remember, read all the directions before you begin, and you will have a better idea of what the bigger picture entails!

Identifying Specific Information from a Printed Communication

Business Memos
Whereas everyday office memos were traditionally typed on paper, photocopied, and distributed, today they are more often typed on computers and distributed via e-mail, both interoffice and externally. Technology has thus made these communications more immediate. It is also helpful for people to read carefully and be familiar with memo components. For example, e-mails automatically provide the same "To:, From:, and Re:" lines traditionally required in paper memos, and in corresponding places—the top of the page/screen. Readers should observe whether "To: names/positions" include all intended recipients in case of misdirection errors or omitted recipients. "From:" informs sender level, role, and who will receive responses when people click "Reply." Users must be careful not to click "Reply All" unintentionally. They should also observe the "CC:" line, typically below "Re:," showing additional recipients.

Classified Ads
Classified advertisements include "Help Wanted" ads informing readers of positions open for hiring, real estate listings, cars for sale, and home and business services available. Traditional ads in newspapers had

to save space, and this necessity has largely transferred to online ads. Because of needing to save space, advertisers employ many abbreviations. For example, here are some examples of abbreviations:

- FT=full-time
- PT=part-time
- A/P=accounts payable
- A/R=accounts receivable
- Asst.=assistant
- Bkkg.=bookkeeping
- Comm.=commission
- Bet.=between
- EOE=equal opportunity employer
- G/L=general ledger
- Immed.=immediately
- Exc.=excellent
- Exp.=experience
- Eves.=evenings
- Secy.=secretary
- Temp=temporary
- Sal=salary
- Req=required
- Refs=references
- Wk=week or work
- WPM=words per minute

Classified ads frequently use abbreviations to take up less space, both on paper and digitally on websites. Those who read these ads will find it less confusing if they learn some common abbreviations used by businesses when advertising job positions. Here are some examples:

- Mgt. = management
- Mgr. = manager
- Mfg. = manufacturing
- Nat'l = national
- Dept. = department
- Min. = minimum
- Yrs. = years
- Nec. = necessary
- Neg. = negotiable
- Oppty = opportunity
- O/T = overtime
- K = 1,000

Readers of classified ads may focus on certain features to the exclusion of others. For example, if a reader sees the job title or salary they are seeking, or notices the experience, education, degree, or other credentials required match their own qualifications perfectly, they may fail to notice other important information, like "No benefits." This is important because the employers are disclosing that they will not provide health insurance, retirement accounts, paid sick leave, paid maternity/paternity leave, paid vacation, etc. to any employee whom they hire. Someone expecting a traditional 9 to 5 job who fails to observe that an ad states "Evenings" or just the abbreviation "Eves" will be disappointed, as will the applicant who overlooks a line saying "Some evenings and weekends reqd." Applicants overlooking information like "Apply in person" may e-mail or mail their resumes and receive no response. The job hopeful with no previous experience and one reference must attend to information like "Minimum 5 yrs. exp, 3 refs," meaning they likely will not qualify.

Employment Ads

Job applicants should pay attention to the information included in classified employment ads. On one hand, they do need to believe and accept certain statements, such as "Please, no phone calls," which is frequently used by employers posting ads on Craigslist and similar websites. New applicants just graduated from or still in college will be glad to see "No exp necessary" in some ads, indicating they need no previous work experience in that job category. "FT/PT" means the employer offers the options of working full-time or part-time, another plus for students. On the other hand, ad readers should also take into consideration the fact that many employers list all the attributes of their *ideal* employee, but they do not necessarily expect to find such a candidate. If a potential applicant's education, training, credentials, and experience are not exactly the same as what the employer lists as desired but are not radically different either, it can be productive to apply anyway, while honestly representing one's actual qualifications.

Atlases

A road atlas is a publication designed to assist travelers who are driving on road trips rather than taking airplanes, trains, ships, etc. Travelers use road atlases to determine which routes to take and places to stop; how to navigate specific cities, locate landmarks, estimate mileages and travel times; see photographs of places they plan to visit; and find other travel-related information. One familiar, reputable road atlas is published by the National Geographic Society. It includes detailed, accurate maps of the United States, Canada, and Mexico; historic sites, scenic routes, recreation information, and points of interest; and its Adventure Edition spotlights 100 top U.S. adventure destinations and most popular national parks. The best-selling road atlas in the United States, also probably the best-known and most trusted, is published annually by Rand McNally, which has published road atlases for many years. It includes maps, mileage charts, information on tourism and road construction, maps of individual city details, and the editor's favorite road trips (in the 2016 edition) including recommended points of interest en route.

Owners' Manuals

An owner's manual is typically a booklet but may also be as short as a page or as long as a book. The purpose of an owner's manual is to give the owner instructions, usually step-by-step, for how to use a specific product or a group or range of products. Manuals accompany consumer products as diverse as cars, computers, tablets, smartphones, printers, home appliances, shop machines, and many others. In addition to directions for operating products, they include important warnings of things *not* to do that pose safety or health hazards or can damage the product and void the manufacturer's product warranty, like immersion in water, exposure to high temperatures, operating something for too long, dropping fragile items, or rough handling. Manuals teach correct operating practices, sequences, precautions, and cautions, averting many costly and/or dangerous mishaps.

Food Labels

When reading the labels on food products, it is often necessary to interpret the nutrition facts and other product information. Without the consumer's being aware and informed of this, much of this information can be very misleading. For example, a popular brand name of corn chips lists the calories, fat, etc. per serving in the nutrition facts printed on the bag, but on closer inspection, it defines a serving size as six chips—far fewer than most people would consume at a time. Serving sizes and the number of servings per container can be unrealistic. For example, a jumbo muffin's wrapper indicates it contains three servings. Most consumers do not divide a muffin and eat only part. It is also rather difficult to cut a muffin into equal thirds.

Product Packaging

Consumers today cannot take product labels at face value. While many people do not read or even look at the information on packages before eating their contents, those who do must use more consideration and analysis than they might expect to understand it. For example, a well-known brand of strawberry-flavored breakfast toaster pastry displays a picture of four whole strawberries on the wrapper. While this looks appealing, encouraging consumers to infer the product contains wholesome fruit—and perhaps even believe it contains four whole strawberries—reading the ingredients list reveals it contains only 2 percent or less of dried strawberries. A consumer must be detail-oriented (and curious or motivated enough) to read the full ingredients list, which also reveals unhealthy corn syrup and high fructose corn syrup high on the list after enriched flour. Consumers must also educate themselves about euphemistically misleading terms: "enriched" flour has vitamins and minerals added, but it is refined flour without whole grain, bran, or fiber.

While manufacturers generally provide extensive information printed on their package labels, it is typically in very small print—many consumers do not read it—and even consumers who do read all the information must look for small details to discover that the information is often not realistic. For example, a box of brownie mix lists grams of fat, total calories, and calories from fat. However, by paying attention to small details like asterisks next to this information, and finding the additional information referenced by the asterisks, the consumer discovers that these amounts are only for the dry mix—not the added eggs, oil, or milk. Consumers typically do not eat dry cake mixes, and having to determine and add the fat and calories from the additional ingredients is inconvenient. In another example, a box of macaroni and cheese mix has an asterisk by the fat grams indicating these are for the macaroni only without the cheese, butter, or milk required, which contributes 6.4 times more fat.

Ingredients' Lists

Consumers can realize the importance of reading drug labeling through an analogy: What might occur if they were to read only part of the directions on a standardized test? Reading only part of the directions on medications can have similar, even more serious consequences. Prescription drug packages typically contain inserts, which provide extremely extensive, thorough, detailed information, including results of clinical trials and statistics showing patient responses and adverse effects. Some over-the-counter medications include inserts, and some do not. "Active ingredients" are those ingredients making medication effective. "Inactive ingredients" including flavorings, preservatives, stabilizers, and emulsifiers have purposes, but not to treat symptoms. "Uses" indicates which symptoms a medication is meant to treat. "Directions" tell the dosage, frequency, maximum daily amount, and other requirements, like "Take with food," "Do not operate heavy machinery while using," etc. Drug labels also state how to store the product, like at what temperature or away from direct sunlight or humidity.

Many drugs which were previously available only by doctor prescription have recently become available over the counter without a prescription. While enough years of testing may have determined that these substances typically do not cause serious problems, consumers must nevertheless thoroughly read and understand all the information on the labels before taking them. If they do not, they could still suffer serious harm. For example, some individuals have allergies to specific substances. Both prescription and over-the-counter medication products list their ingredients, including warnings about allergies. Allergic reactions can include anaphylactic shock, which can be fatal if not treated immediately. Also, consumers must read and follow dosing directions: taking more than directed can cause harm, and taking less can be ineffective to treat symptoms. Some medication labels warn not to mix them with certain other drugs to avoid harmful drug interactions. Additionally, without reading ingredients, some consumers take multiple products including the same active ingredients, resulting in overdoses.

Identifying Information from a Graphic Representation of Information

Line Graphs

Line graphs are useful for visually representing data that vary continuously over time, like an individual student's test scores. The horizontal or *x*-axis shows dates/times; the vertical or y-axis shows point values. A dot is plotted on the point where each horizontal date line intersects each vertical number line, and then these dots are connected, forming a line. Line graphs show whether changes in values over time exhibit trends like ascending, descending, flat, or more variable, like going up and down at different times. For example, suppose a student's scores on the same type of reading test were 75% in October, 80% in November, 78% in December, 82% in January, 85% in February, 88% in March, and 90% in April. A line graph of these scores would look like this.

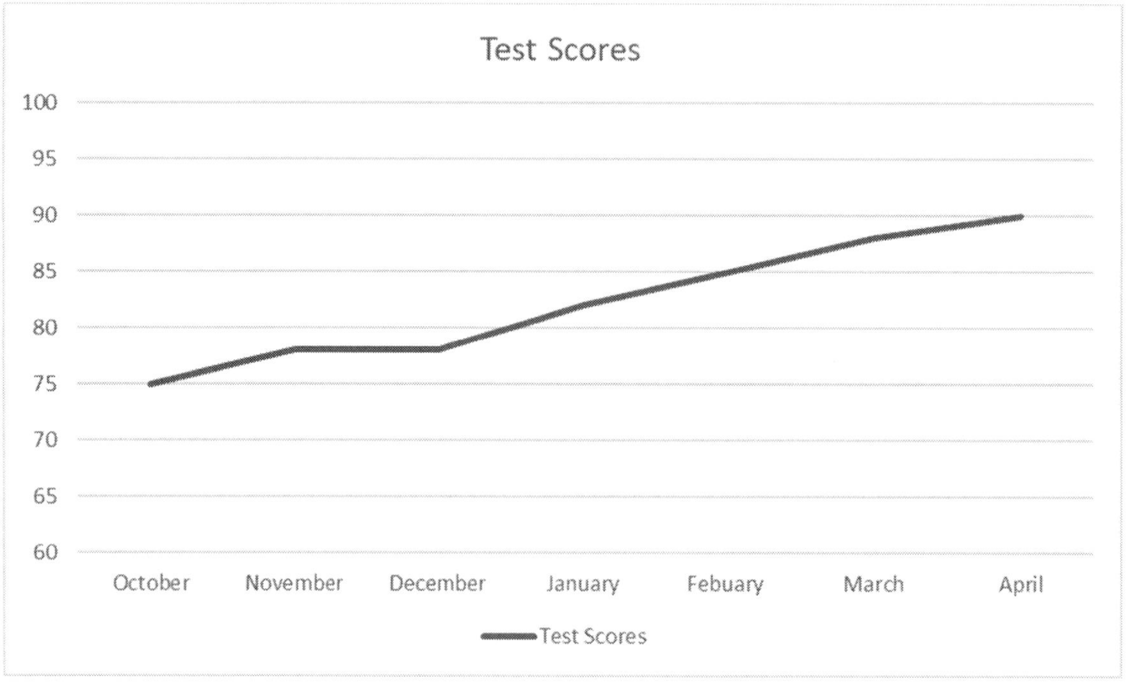

Bar Graphs

Bar graphs feature equally spaced, horizontal or vertical rectangular bars representing numerical values. They can show change over time as line graphs do, but unlike line graphs, bar graphs can also show differences and similarities among values at a single point in time. Bar graphs are also helpful for visually representing data from different categories, especially when the horizontal axis displays some value that is not numerical, like various countries with inches of annual rainfall.

The following is a bar graph that compares different classes and how many books they read:

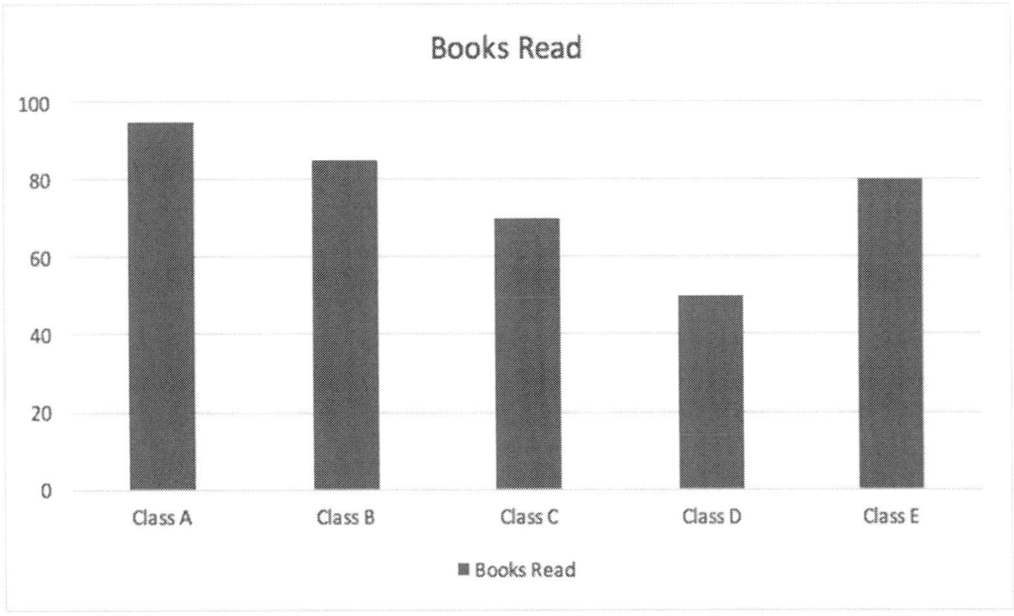

Pie Charts

Pie charts, also called circle graphs, are good for representing percentages or proportions of a whole quantity because they represent the whole as a circle or "pie," with the various proportion values shown as "slices" or wedges of the pie. This gives viewers a clear idea of how much of a total each item occupies. To calculate central angles to make each portion the correct size, multiply each percentage by 3.6 (= 360/100). For example, biologists may have information that 60% of Americans have brown eyes, 20% have hazel eyes, 15% have blue eyes, and 5% have green eyes. A pie chart of these distributions would look like this:

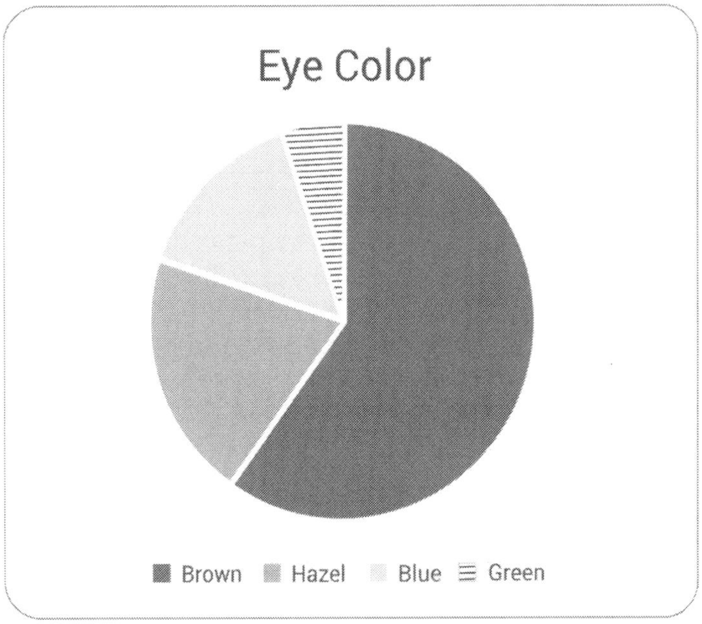

Line Plots

Rather than showing trends or changes over time like line graphs, line plots show the frequency with which a value occurs in a group. Line plots are used for visually representing data sets that total 50 or fewer values. They make visible features like gaps between some data points, clusters of certain numbers/number ranges, and outliers (data points with significantly smaller or larger values than others). For example, the age ranges in a class of nursing students might appear like this in a line plot:

xxxxxxxxxx	xxxxxx	xx	x	xxx	xx	x
18	23	28	33	38	43	48

Pictograms

Magazines, newspapers, and other similar publications designed for consumption by the general public often use pictograms to represent data. **Pictograms** feature icons or symbols that look like whatever category of data is being counted, like little silhouettes shaped like human beings commonly used to represent people. If the data involve large numbers, like populations, one person symbol might represent one million people, or one thousand, etc. For smaller values, such as how many individuals out of ten fit a given description, one symbol might equal one person. Male and female silhouettes are used to differentiate gender, and child shapes for children. Little clock symbols are used to represent amounts of time, such as a given number of hours; calendar pages might depict months; suns and moons could show days and nights; hourglasses might represent minutes. While pictogram symbols are easily recognizable and appealing to general viewers, one disadvantage is that it is difficult to display partial symbols for in-between quantities.

Recognizing Events in a Sequence

Sequence structure is the order of events in which a story or information is presented to the audience. Sometimes the text will be presented in chronological order, or sometimes it will be presented by displaying the most recent information first, then moving backwards in time. The sequence structure depends on the author, the context, and the audience. The structure of a text also depends on the genre in which the text is written. Is it literary fiction? Is it a magazine article? Is it instructions for how to complete a certain task? Different genres will have different purposes for switching up the sequence.

Narrative Structure

The structure presented in literary fiction, called **narrative structure**, is the foundation on which the text moves. The narrative structure comes from the plot and setting. The plot is the sequence of events in the narrative that move the text forward through cause and effect. The setting is the place or time period in which the story takes place. Narrative structure has two main categories: linear and nonlinear.

Linear Narrative

A narrative is linear when it is told in chronological order. Traditional linear narratives will follow the plot diagram below depicting the narrative arc. The narrative arc consists of the exposition, conflict, rising action, climax, falling action, and resolution.

- Exposition: The exposition is in the beginning of a narrative and introduces the characters, setting, and background information of the story. The exposition provides the context for the upcoming narrative. Exposition literally means "a showing forth" in Latin.

- Conflict: In a traditional narrative, the conflict appears toward the beginning of the story after the audience becomes familiar with the characters and setting. The conflict is a single instance between characters, nature, or the self, in which the central character is forced to make a decision or move forward with some kind of action. The conflict presents something for the main character, or protagonist, to overcome.

- Rising Action: The rising action is the part of the story that leads into the climax. The rising action will feature the development of characters and plot and creates the tension and suspense that eventually lead to the climax.

- Climax: The climax is the part of the story where the tension produced in the rising action comes to a culmination. The climax is the peak of the story. In a traditional structure, everything before the climax builds up to it, and everything after the climax falls from it. It is the height of the narrative and is usually either the most exciting part of the story or is marked by some turning point in the character's journey.

- Falling Action: The falling action happens as a result of the climax. Characters continue to develop, although there is a wrapping up of loose ends here. The falling action leads to the resolution.

- Resolution: The resolution is where the story comes to an end and usually leaves the reader with the satisfaction of knowing what happened within the story and why. However, stories do not always end in this fashion. Sometimes readers can be confused or frustrated at the end from lack of information or the absence of a happy ending.

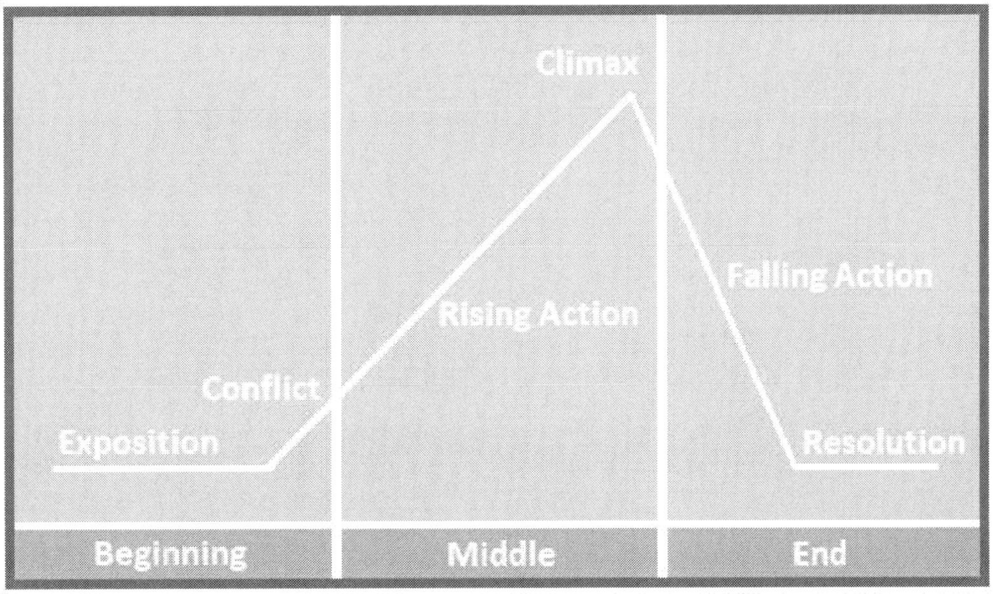

Nonlinear Narrative

A **nonlinear narrative** deviates from the traditional narrative because it does not always follow the traditional plot structure of the narrative arc. Nonlinear narratives may include structures that are disjointed, circular, or disruptive, in the sense that they do not follow chronological order. **In medias res** is an example of a nonlinear structure. *In medias res* is Latin for "in the middle of things," which is how many ancient texts, especially epic poems, began their story, such as Homer's *Iliad.* Instead of having a clear exposition with a full development of characters, they would begin right in the middle of the action.

Many modernist texts in the late nineteenth and early twentieth centuries experimented with disjointed narratives, moving away from traditional linear narrative. Disjointed narratives are depicted in novels like *Catch 22*, where the author, Joseph Heller, structures the narrative based on free association of ideas rather than chronology. Another nonlinear narrative can be seen in the novel *Wuthering Heights*, written by Emily Brontë; after the first chapter, the narrative progresses retrospectively instead of chronologically. There seem to be two narratives in *Wuthering Heights* working at the same time: a present narrative as well as a past narrative. Authors employ disrupting narratives for various reasons; some use it for the purpose of creating situational irony for the readers, while some use it to create a certain effect, such as excitement, discomfort, or fear.

Sequence Structure in Technical Documents

The purpose of technical documents, such as instruction manuals, cookbooks, or "user-friendly" documents, is to provide information to users as clearly and efficiently as possible. In order to do this, the sequence structure in technical documents should be as straightforward as possible. This usually involves some kind of chronological order or a direct sequence of events. For example, someone who is reading an instruction manual on how to set up their Smart TV wants directions in a clear, simple, straightforward manner that does not confuse them or leave them guessing about the proper sequence.

Sequence Structure in Informational Texts

The structure of informational texts depends on the specific genre. For example, a newspaper article may start by stating an exciting event that happened, then talk about that event in chronological order. Many informational texts also use **cause and effect structure**, which describes an event and then identifies reasons for why that event occurred. Some essays may write about their subjects by way of **comparison and contrast**, which is a structure that compares two things or contrasts them to highlight their differences. Other documents, such as proposals, will have a **problem to solution structure**, where the document highlights some kind of problem and then offers a solution. Finally, some informational texts are written with lush details and description in order to captivate the audience, allowing them to visualize the information presented to them. This type of structure is known as **descriptive.**

Recognizing the Structure of Texts in Various Formats

Text structure is the way in which the author organizes and presents textual information so readers can follow and comprehend it. One kind of text structure is sequence. This means the author arranges the text in a logical order from beginning to middle to end. There are three types of sequences:

- Chronological: ordering events in time from earliest to latest
- Spatial: describing objects, people, or spaces according to their relationships to one another in space
- Order of Importance: addressing topics, characters, or ideas according to how important they are, from either least important to most important

Chronological sequence is the most common sequential text structure. Readers can identify sequential structure by looking for words that signal it, like *first, earlier, meanwhile, next, then, later, finally;* and specific times and dates the author includes as chronological references.

Problem-Solution Text Structure
The problem-solution text structure organizes textual information by presenting readers with a problem and then developing its solution throughout the course of the text. The author may present a variety of alternatives as possible solutions, eliminating each as they are found unsuccessful, or gradually leading up to the ultimate solution. For example, in fiction, an author might write a murder mystery novel and have the character(s) solve it through investigating various clues or character alibis until the killer is identified. In nonfiction, an author writing an essay or book on a real-world problem might discuss various alternatives and explain their disadvantages or why they would not work before identifying the best solution. For scientific research, an author reporting and discussing scientific experiment results would explain why various alternatives failed or succeeded.

Comparison-Contrast Text Structure
Comparison identifies similarities between two or more things. **Contrast** identifies differences between two or more things. Authors typically employ both to illustrate relationships between things by highlighting their commonalities and deviations. For example, a writer might compare Windows and Linux as operating systems, and contrast Linux as free and open-source vs. Windows as proprietary. When writing an essay, sometimes it is useful to create an image of the two objects or events you are comparing or contrasting. Venn diagrams are useful because they show the differences as well as the similarities between two things. Once you've seen the similarities and differences on paper, it might be helpful to create an outline of the essay with both comparison and contrast. Every outline will look different because every two or more things will have a different number of comparisons and contrasts. Say you are trying to compare and contrast carrots with sweet potatoes. Here is an example of a compare/contrast outline using those topics:

- Introduction: Talk about why you are comparing and contrasting carrots and sweet potatoes. Give the thesis statement.
- Body paragraph 1: Sweet potatoes and carrots are both root vegetables (similarity)
- Body paragraph 2: Sweet potatoes and carrots are both orange (similarity)
- Body paragraph 3: Sweet potatoes and carrots have different nutritional components (difference)
- Conclusion: Restate the purpose of your comparison/contrast essay.

Of course, if there is only one similarity between your topics and two differences, you will want to rearrange your outline. Always tailor your essay to what works best with your topic.

Descriptive Text Structure

Description can be both a type of text structure and a type of text. Some texts are descriptive throughout entire books. For example, a book may describe the geography of a certain country, state, or region, or tell readers all about dolphins by describing many of their characteristics. Many other texts are not descriptive throughout but use descriptive passages within the overall text. The following are a few examples of descriptive text:

- When the author describes a character in a novel
- When the author sets the scene for an event by describing the setting
- When a biographer describes the personality and behaviors of a real-life individual
- When a historian describes the details of a particular battle within a book about a specific war
- When a travel writer describes the climate, people, foods, and/or customs of a certain place

A hallmark of description is using sensory details, painting a vivid picture so readers can imagine it almost as if they were experiencing it personally.

Cause and Effect Text Structure

When using cause and effect to extrapolate meaning from text, readers must determine the cause when the author only communicates effects. For example, if a description of a child eating an ice cream cone includes details like beads of sweat forming on the child's face and the ice cream dripping down her hand faster than she can lick it off, the reader can infer or conclude it must be hot outside. A useful technique for making such decisions is wording them in "If...then" form, e.g. "If the child is perspiring and the ice cream melting, it may be a hot day." Cause and effect text structures explain why certain events or actions resulted in particular outcomes. For example, an author might describe America's historical large flocks of dodo birds, the fact that gunshots did not startle/frighten dodos, and that because dodos did not flee, settlers killed whole flocks in one hunting session, explaining how the dodo was hunted into extinction.

Integrating Data from Multiple Sources in Various Formats, Including Media

Books as Resources

When a student has an assignment to research and write a paper, one of the first steps after determining the topic is to select research sources. The student may begin by conducting an Internet or library search of the topic, may refer to a reading list provided by the instructor, or may use an annotated bibliography of works related to the topic. To evaluate the worth of the book for the research paper, the student first considers the book title to get an idea of its content. Then the student can scan the book's table of contents for chapter titles and topics to get further ideas of their applicability to the topic. The student may also turn to the end of the book to look for an alphabetized index. Most academic textbooks and scholarly works have these; students can look up key topic terms to see how many are included and how many pages are devoted to them.

Journal Articles

Like books, **journal articles** are primary or secondary sources the student may need to use for researching any topic. To assess whether a journal article will be a useful source for a particular paper topic, a student can first get some idea about the content of the article by reading its title and subtitle, if any exists. Many journal articles, particularly scientific ones, include abstracts. These are brief summaries of the content. The student should read the abstract to get a more specific idea of whether the experiment, literature review, or other work documented is applicable to the paper topic. Students should also check the references at the end of the article, which today often contain links to related works for exploring the topic further.

Encyclopedias and Dictionaries

Dictionaries and **encyclopedias** are both reference books for looking up information alphabetically. Dictionaries are more exclusively focused on vocabulary words. They include each word's correct spelling, pronunciation, variants, part(s) of speech, definitions of one or more meanings, and examples used in a sentence. Some dictionaries provide illustrations of certain words when these inform the meaning. Some dictionaries also offer synonyms, antonyms, and related words under a word's entry. Encyclopedias, like dictionaries, often provide word pronunciations and definitions. However, they have broader scopes: one can look up entire subjects in encyclopedias, not just words, and find comprehensive, detailed information about historical events, famous people, countries, disciplines of study, and many other things. Dictionaries are for finding word meanings, pronunciations, and spellings; encyclopedias are for finding breadth and depth of information on a variety of topics.

Card Catalogs

A **card catalog** is a means of organizing, classifying, and locating the large numbers of books found in libraries. Without being able to look up books in library card catalogs, it would be virtually impossible to find them on library shelves. Card catalogs may be on traditional paper cards filed in drawers, or electronic catalogs accessible online; some libraries combine both. Books are shelved by subject area; subjects are coded using formal classification systems—standardized sets of rules for identifying and labeling books by subject and author. These assign each book a call number: a code indicating the classification system, subject, author, and title. Call numbers also function as bookshelf "addresses" where books can be located. Most public libraries use the Dewey Decimal Classification System. Most university, college, and research libraries use the Library of Congress Classification. Nursing students will also encounter the National Institute of Health's National Library of Medicine Classification System, which major collections of health sciences publications utilize.

Databases

A **database** is a collection of digital information organized for easy access, updating, and management. Users can sort and search databases for information. One way of classifying databases is by content, i.e. full-text, numerical, bibliographical, or images. Another classification method used in computing is by organizational approach. The most common approach is a relational database, which is tabular and defines data so they can be accessed and reorganized in various ways. A distributed database can be reproduced or interspersed among different locations within a network. An object-oriented database is organized to be aligned with object classes and subclasses defining the data. Databases usually collect files like product inventories, catalogs, customer profiles, sales transactions, student bodies, and resources. An associated set of application programs is a database management system or database manager. It enables users to specify which reports to generate, control access to reading and writing data, and analyze database usage. Structured Query Language (SQL) is a standard computer language for updating, querying, and otherwise interfacing with databases.

Words in Context

Context Clues

Readers can often figure out what unfamiliar words mean without interrupting their reading to look them up in dictionaries by examining context. **Context** includes the other words or sentences in a passage. One common context clue is the root word and any affixes (prefixes/suffixes). Another common context clue is a synonym or definition included in the sentence. Sometimes both exist in the same sentence. Here's an example:

> Scientists who study birds are *ornithologists*.

Many readers may not know the word *ornithologist*. However, the example contains a definition (scientists who study birds). The reader may also have the ability to analyze the suffix (*-logy*, meaning the study of) and root (*ornitho-*, meaning bird).

Another common context clue is a sentence that shows differences. Here's an example:

> Birds *incubate* their eggs outside of their bodies, unlike mammals.

Some readers may be unfamiliar with the word *incubate*. However, since the sentence includes the phrase "unlike mammals," the reader can infer that incubate relates to an aspect of prenatal development that mammals and birds do not have in common, such as the way they keep the embryo at a temperature suitable for development.

In addition to analyzing the etymology of a word's root and affixes and extrapolating word meaning from sentences that contrast an unknown word with an antonym, readers can also determine word meanings from sentence context clues based on logic. Here's an example:

> Birds are always looking out for predators that could attack their young.

The reader who is unfamiliar with the word *predator* could determine from the context of the sentence that predators usually prey upon baby birds and possibly other young animals. Readers might also use the context clue of etymology here, as *predator* and *prey* have the same root.

Analyzing Word Parts

Readers can learn some etymologies, or origins, of words and their parts, making it easier to break down new words into components and analyze their combined meanings. For example, the root word soph is Greek for "wise" or "knowledge." Knowing this informs the meanings of English words including *sophomore, sophisticated,* and *philosophy.* Those who also know that phil is Greek for "love" will realize that philosophy means "love of knowledge." They can then extend this knowledge of *phil* to understand *philanthropist* (one who loves people), *bibliophile* (book lover), *philharmonic* (loving harmony), *hydrophilic* (water-loving), and so on. In addition, *phob* derives from the Greek *phobos*, meaning "fear." Words with this root indicate fear of various things *acrophobia* (fear of heights), *arachnophobia* (fear of spiders), *claustrophobia* (fear of enclosed spaces), *ergophobia* (fear of work), and *hydrophobia* (fear of water), among others.

Some English word origins from other languages, like ancient Greek, are found in large numbers and varieties of English words. An advantage of the shared ancestry of these words is that once readers recognize the meanings of some Greek words or word roots, they can determine or at least get an idea of what many different English words mean. As an example, the Greek word *métron* means to measure, a measure, or something used to measure; the English word meter derives from it. Knowing this informs many other English words, including *altimeter, barometer, diameter, hexameter, isometric,* and *metric.* While readers must know the meanings of the other parts of these words to decipher their meaning fully, they already have an idea that they are all related in some way to measures or measuring.

While all English words ultimately derive from a proto-language known as Indo-European, many of them historically came into the developing English vocabulary later, from sources like the ancient Greeks, the Latin used throughout Europe and much of the Middle East during the reign of the Roman Empire, and the Anglo-Saxon languages used by England's early tribes. In addition to classic revivals and native foundations, by the Renaissance era other influences included French, German, Italian, and Spanish. Today we can often discern English word meanings by knowing common roots and affixes, particularly from Greek and Latin.

The following is a list of common prefixes and their meanings:

Prefix	Definition	Examples
a-	without	atheist, agnostic
ad-	to, toward	advance
ante-	before	antecedent, antedate
anti-	opposing	antipathy, antidote
auto-	self	autonomy, autobiography
bene-	well, good	benefit, benefactor
bi-	two	bisect, biennial
bio-	life	biology, biosphere
chron-	time	chronometer, synchronize
circum-	around	circumspect, circumference
com-	with, together	commotion, complicate
contra-	against, opposing	contradict, contravene
cred-	belief, trust	credible, credit
de-	from	depart
dem-	people	demographics, democracy
dis-	away, off, down, not	dissent, disappear
equi-	equal, equally	equivalent
ex-	from, out of	extract
for-	away, off, from	forget, forswear
fore-	before, previous	foretell, forefathers
homo-	same, equal	homogenized
hyper-	excessive, over	hypercritical, hypertension
in-	in, into	intrude, invade
inter-	among, between	intercede, interrupt
mal-	bad, poorly, not	malfunction
micr(o)-	small	microbe, microscope
mis-	bad, poorly, not	misspell, misfire
mono-	one, single	monogamy, monologue
mor-	die, death	mortality, mortuary
neo-	new	neolithic, neoconservative
non-	not	nonentity, nonsense
omni-	all, everywhere	omniscient
over-	above	overbearing
pan-	all, entire	panorama, pandemonium
para-	beside, beyond	parallel, paradox

phil-	love, affection	philosophy, philanthropic
poly-	many	polymorphous, polygamous
pre-	before, previous	prevent, preclude
prim-	first, early	primitive, primary
pro-	forward, in place of	propel, pronoun
re-	back, backward, again	revoke, recur
sub-	under, beneath	subjugate, substitute
super-	above, extra	supersede, supernumerary
trans-	across, beyond, over	transact, transport
ultra-	beyond, excessively	ultramodern, ultrasonic, ultraviolet
un-	not, reverse of	unhappy, unlock
vis-	to see	visage, visible

The following is a list of common suffixes and their meanings:

Suffix	**Definition**	**Examples**
-able	likely, able to	capable, tolerable
-ance	act, condition	acceptance, vigilance
-ard	one that does excessively	drunkard, wizard
-ation	action, state	occupation, starvation
-cy	state, condition	accuracy, captaincy
-er	one who does	teacher
-esce	become, grow, continue	convalesce, acquiesce
-esque	in the style of, like	picturesque, grotesque
-ess	feminine	waitress, lioness
-ful	full of, marked by	thankful, zestful
-ible	able, fit	edible, possible, divisible
-ion	action, result, state	union, fusion
-ish	suggesting, like	churlish, childish
-ism	act, manner, doctrine	barbarism, socialism
-ist	doer, believer	monopolist, socialist
-ition	action, result, state,	sedition, expedition
-ity	quality, condition	acidity, civility
-ize	cause to be, treat with	sterilize, mechanize, criticize
-less	lacking, without	hopeless, countless
-like	like, similar	childlike, dreamlike
-ly	like, of the nature of	friendly, positively
-ment	means, result, action	refreshment, disappointment
-ness	quality, state	greatness, tallness
-or	doer, office, action	juror, elevator, honor
-ous	marked by, given to	religious, riotous
-some	apt to, showing	tiresome, lonesome
-th	act, state, quality	warmth, width
-ty	quality, state	enmity, activity

Interpreting the Meaning of Words and Phrases Using Context

When readers encounter an unfamiliar word in text, they can use the surrounding context to help determine the word's meaning. The text's overall subject matter, the specific chapter or section, and the immediate sentence context can all provide clues to help the reader understand the word. Among others, one category of context clues is grammar. For example, the position of a word in a sentence and its relationship to the other words can help the reader establish whether the unfamiliar word is a verb, a noun, an adjective, an adverb, etc. This narrows down the possible meanings of the word to one part of speech. However, this may be insufficient. In the sentence, "Many birds migrate twice yearly," the reader can determine that the italicized word is a verb. While it probably does not mean eat or drink (because birds would need to do those actions more than twice each year), it could mean travel, mate, lay eggs, hatch, molt, etc.

Some words can have a number of different meanings depending on how they are used. For example, the word *fly* has a different meaning in each of the following sentences:

- "His trousers have a fly on them."
- "He swatted the fly on his trousers."
- "Those are some fly trousers."
- "They went fly fishing."
- "She hates to fly."
- "If humans were meant to fly, they would have wings."

As strategies, readers can try substituting a familiar word for an unfamiliar one and see whether it makes sense in the sentence. They can also identify other words in a sentence, offering clues to an unfamiliar word's meaning.

Distinguishing Between Fact and Opinion, Biases, and Stereotypes

Facts and Opinions

A **fact** is a statement that is true empirically or an event that has actually occurred in reality and can be proven or supported by evidence; it is generally objective. In contrast, an **opinion** is subjective, representing something that someone believes rather than something that exists in the absolute. People's individual understandings, feelings, and perspectives contribute to variations in opinion. Though facts are typically objective in nature, in some instances, a statement of fact may be both factual and yet also subjective. For example, emotions are individual subjective experiences. If an individual says that they feel happy or sad, the feeling is subjective, but the statement is factual; hence, it is a subjective fact. In contrast, if one person tells another that the other is feeling happy or sad—whether this is true or not—that is an assumption or an opinion.

Biases

Biases usually occur when someone allows their personal preferences or ideologies to interfere with what should be an objective decision. In personal situations, someone is biased towards someone if they favor them in an unfair way. In academic writing, being biased in your sources means leaving out objective information that would turn the argument one way or the other. The evidence of bias in academic writing makes the text less credible, so be sure to present all viewpoints when writing, not just your own, so to avoid coming off as biased. Being objective when presenting information or dealing with people usually allows the person to gain more credibility.

Stereotypes

Stereotypes are preconceived notions that place a particular rule or characteristics on an entire group of people. Stereotypes are usually offensive to the group they refer to or allies of that group, and often have negative connotations. The reinforcement of stereotypes isn't always obvious. Sometimes stereotypes can be very subtle and are still widely used in order for people to understand categories within the world. For example, saying that women are more emotional and intuitive than men is a stereotype, although this is still an assumption used by many in order to understand the differences between one another.

Denotation and Connotation

Denotation, a word's explicit definition, is often set in comparison to connotation, the emotional, cultural, social, or personal implications associated with a word. Denotation is more of an objective definition, whereas connotation can be more subjective, although many connotative meanings of words are similar for certain cultures. The denotative meanings of words are usually based on facts, and the connotative meanings of words are usually based on emotion. Here are some examples of words and their denotative and connotative meanings in Western culture:

Word	Denotative Meaning	Connotative Meaning
Home	A permanent place where one lives, usually as a member of a family.	A place of warmth; a place of familiarity; comforting; a place of safety and security. "Home" usually has a positive connotation.
Snake	A long reptile with no limbs and strong jaws that moves along the ground; some snakes have a poisonous bite.	An evil omen; a slithery creature (human or nonhuman) that is deceitful or unwelcome. "Snake" usually has a negative connotation.
Winter	A season of the year that is the coldest, usually from December to February in the northern hemisphere and from June to August in the southern hemisphere.	Circle of life, especially that of death and dying; cold or icy; dark and gloomy; hibernation, sleep, or rest. Winter can have a negative connotation, although many who have access to heat may enjoy the snowy season from their homes.

Evaluating the Author's Purpose in a Given Text

Authors may have many purposes for writing a specific text. They could be imparting information, entertaining their audience, expressing their own feelings, or trying to persuade their readers of a particular position. Authors' purposes are their reasons for writing something. A single author may have one overriding purpose for writing or multiple reasons. An author may explicitly state their intention in the text, or the reader may need to infer that intention. When readers can identify the author's purpose, they are better able to analyze information in the text. By knowing why the author wrote the text, readers can glean ideas for how to approach it. The following is a list of questions readers can ask in order to discern an author's purpose for writing a text:

- Does the title of the text give you any clues about its purpose?
- Was the purpose of the text to give information to readers?
- Did the author want to describe an event, issue, or individual?
- Was it written to express emotions and thoughts?
- Did the author want to convince readers to consider a particular issue?
- Do you think the author's primary purpose was to entertain?
- Why do you think the author wrote this text from a certain point of view?
- What is your response to the text as a reader?
- Did the author state their purpose for writing it?

Rather than simply consuming the text, readers should attempt to interpret the information being presented. Being able to identify an author's purpose efficiently improves reading comprehension, develops critical thinking, and makes students more likely to consider issues in depth before accepting writer viewpoints. Authors of fiction frequently write to entertain readers. Another purpose for writing fiction is making a political statement; for example, Jonathan Swift wrote "A Modest Proposal" (1729) as a political satire. Another purpose for writing fiction as well as nonfiction is to persuade readers to take some action or further a particular cause. Fiction authors and poets both frequently write to evoke certain moods; for example, Edgar Allan Poe wrote novels, short stories, and poems that evoke moods of gloom, guilt, terror, and dread. Another purpose of poets is evoking certain emotions: love is popular, as in Shakespeare's sonnets and numerous others. In "The Waste Land" (1922), T.S. Eliot evokes society's alienation, disaffection, sterility, and fragmentation.

Authors seldom directly state their purposes in texts. Some students may be confronted with nonfiction texts such as biographies, histories, magazine and newspaper articles, and instruction manuals, among others. To identify the purpose in nonfiction texts, students can ask the following questions:

- Is the author trying to teach something?
- Is the author trying to persuade the reader?
- Is the author imparting factual information only?
- Is this a reliable source?
- Does the author have some kind of hidden agenda?

To apply author purpose in nonfictional passages, students can also analyze sentence structure, word choice, and transitions to answer the aforementioned questions and to make inferences. For example, authors wanting to convince readers to view a topic negatively often choose words with negative connotations.

Narrative Writing

Narrative writing tells a story. The most prominent type of narrative writing is the fictional novel. Here are some examples:

- Mark Twain's *The Adventures of Tom Sawyer and The Adventures of Huckleberry Finn*
- Victor Hugo's *Les Misérables*
- Charles Dickens' *Great Expectations*, *David Copperfield*, and *A Tale of Two Cities*
- Jane Austen's *Northanger Abbey*, *Mansfield Park*, *Pride and Prejudice*, *Sense and Sensibility*, and *Emma*

- Toni Morrison's *Beloved*, *The Bluest Eye*, and *Song of Solomon*
- Gabriel García Márquez's *One Hundred Years of Solitude* and *Love in the Time of Cholera*

Nonfiction works can also appear in narrative form. For example, some authors choose a narrative style to convey factual information about a topic, such as a specific animal, country, geographic region, and scientific or natural phenomenon.

Narrative writing tells a story, and the one telling the story is called the narrator. The narrator may be a fictional character telling the story from their own viewpoint. This narrator uses the first person (*I, me, my, mine* and *we, us, our,* and *ours*). The narrator may also be the author; for example, when Louisa May Alcott writes "Dear reader" in *Little Women*, she (the author) addresses us as readers. In this case, the novel is typically told in third person, referring to the characters as he, she, they, or them. Another more common technique is the omniscient narrator; in other words, the story is told by an unidentified individual who sees and knows everything about the events and characters—not only their externalized actions, but also their internalized feelings and thoughts. Second person narration, which addresses readers as you throughout the text, is more uncommon than the first and third person options.

Expository Writing

Expository writing is also known as informational writing. Its purpose is not to tell a story as in narrative writing, to paint a picture as in descriptive writing, or to persuade readers to agree with something as in argumentative writing. Rather, its point is to communicate information to the reader. As such, the point of view of the author will necessarily be more objective. Expository writing does not appeal to emotions or reason, nor does it use subjective descriptions to sway the reader's opinion or thinking; rather, expository writing simply provides facts, evidence, observations, and objective descriptions of the subject matter. Some examples of expository writing include research reports, journal articles, books about history, academic textbooks, essays, how-to articles, user instruction manuals, news articles, and other factual journalistic reports.

Technical Writing

Technical writing is similar to expository writing because it provides factual and objective information. Indeed, it may even be considered a subcategory of expository writing. However, technical writing differs from expository writing in two ways: (1) it is specific to a particular field, discipline, or subject, and (2) it uses technical terminology that belongs only to that area. Writing that uses technical terms is intended only for an audience familiar with those terms. An example of technical writing would be a manual on computer programming and use.

Persuasive Writing

Persuasive writing or ***argumentative writing*** attempts to convince the reader to agree with the author's position. Some writers may respond to other writers' arguments by making reference to those authors or texts and then disagreeing with them. However, another common technique is for the author to anticipate opposing viewpoints, both from other authors and from readers. The author brings up these opposing viewpoints, and then refutes them before they can even be raised, strengthening the author's argument. Writers persuade readers by appealing to the readers' reason and emotion, as well as to their own character and credibility. Aristotle called these appeals ***logos***, ***pathos***, and ***ethos***, respectively.

Evaluating the Author's Point of View in a Given Text

When a writer tells a story using the first person, readers can identify this by the use of first-person pronouns, like *I, me, we, us,* etc. However, first-person narratives can be told by different people or from different points of view. For example, some authors write in the first person to tell the story from the main character's viewpoint, as Charles Dickens did in his novels *David Copperfield* and *Great Expectations.* Some authors write in the first person from the viewpoint of a fictional character in the story, but not necessarily the main character. For example, F. Scott Fitzgerald wrote *The Great Gatsby* as narrated by Nick Carraway, a character in the story, about the main characters, Jay Gatsby and Daisy Buchanan. Other authors write in the first person, but as the omniscient narrator—an often unnamed person who knows all of the characters' inner thoughts and feelings. Writing in first person as oneself is more common in nonfiction.

Third Person
The third-person narrative is probably the most prevalent voice used in fictional literature. While some authors tell stories from the point of view and in the voice of a fictional character using the first person, it is a more common practice to describe the actions, thoughts, and feelings of fictional characters in the third person using *he, him, she, her, they, them,* etc.

Although plot and character development are both necessary and possible when writing narrative from a first-person point of view, they are also more difficult, particularly for new writers and those who find it unnatural or uncomfortable to write from that perspective. Therefore, writing experts advise beginning writers to start out writing in the third person. A big advantage of third-person narration is that the writer can describe the thoughts, feelings, and motivations of every character in a story, which is not possible for the first-person narrator. Third-person narrative can impart information to readers that the characters do not know. On the other hand, beginning writers often regard using the third-person point of view as more difficult because they must write about the feelings and thoughts of every character, rather than only about those of the protagonist.

Second Person
Narrative written in the second person addresses someone else as "you." In novels and other fictional works, the second person is the narrative voice most seldom used. The primary reason for this is that it often reads in an awkward manner, which prevents readers from being drawn into the fictional world of the novel. The second person is more often used in informational text, especially in how-to manuals, guides, and other instructions.

First Person
First person uses pronouns such as *I, me, we, my, us, and our.* Some writers naturally find it easier to tell stories from their own points of view, so writing in the first person offers advantages for them. The first-person voice is better for interpreting the world from a single viewpoint, and for enabling reader immersion in one protagonist's experiences. However, others find it difficult to use the first-person narrative voice. Its disadvantages can include overlooking the emotions of characters, forgetting to include description, producing stilted writing, using too many sentence structures involving "I did . . .", and not devoting enough attention to the story's "here-and-now" immediacy.

Using Text Features

Table of Contents and Index
When examining a book, a journal article, a monograph, or other publication, the table of contents is in the front. In books, it is typically found following the title page, publication information (often on the facing side of the title page), and dedication page, when one is included. In shorter publications, the table of contents may follow the title page, or the title on the same page. The **table of contents** in a book lists the number and title of each chapter and its beginning page number. An **index**, which is most common in books but may also be included in shorter works, is at the back of the publication. Books, especially academic texts, frequently have two: a subject index and an author index. Readers can look alphabetically

for specific subjects in the subject index. Likewise, they can look up specific authors cited, quoted, discussed, or mentioned in the author index.

The index in a book offers particular advantages to students. For example, college course instructors typically assign certain textbooks, but do not expect students to read the entire book from cover to cover immediately. They usually assign specific chapters to read in preparation for specific lectures and/or discussions in certain upcoming classes. Reading portions at a time, some students may find references they either do not fully understand or want to know more about. They can look these topics up in the book's subject index to find them in later chapters. When a text author refers to another author, students can also look up the name in the book's author index to find all page numbers of all other references to that author. College students also typically are assigned research papers to write. A book's subject and author indexes can guide students to pages that may help inform them of other books to use for researching paper topics.

Headings

Headings and subheadings concisely inform readers what each section of a paper contains, as well as showing how its information is organized both visually and verbally. Headings are typically up to about five words long. They are not meant to give in-depth analytical information about the topic of their section, but rather an idea of its subject matter. Text authors should maintain consistent style across all headings. Readers should not expect headings if there is not material for more than one heading at each level, just as a list is unnecessary for a single item. Subheadings may be a bit longer than headings because they expand upon them. Readers should skim the subheadings in a paper to use them as a map of how content is arranged. Subheadings are in smaller fonts than headings to mirror relative importance. Subheadings are not necessary for every paragraph. They should enhance content, not substitute for topic sentences.

When a heading is brief, simple, and written in the form of a question, it can have the effect of further drawing readers into the text. An effective author will also answer the question in the heading soon in the following text. Question headings and their text answers are particularly helpful for engaging readers with average reading skills. Both headings and subheadings are most effective with more readers when they are obvious, simple, and get to their points immediately. Simple headings attract readers; simple subheadings allow readers a break, during which they also inform reader decisions whether to continue reading or not. Headings stand out from other text through boldface, but also italicizing and underlining them would be excessive. Uppercase-lowercase headings are easier for readers to comprehend than all capitals. More legible fonts are better. Some experts prefer serif fonts in text, but sans-serif fonts in headings. Brief subheadings that preview upcoming chunks of information reach more readers.

Text Features

Textbooks that are designed well employ varied text features for organizing their main ideas, illustrating central concepts, spotlighting significant details, and signaling evidence that supports the ideas and points conveyed. When a textbook uses these features in recurrent patterns that are predictable, it makes it easier for readers to locate information and come up with connections. When readers comprehend how to make use of text features, they will take less time and effort deciphering how the text is organized, leaving them more time and energy for focusing on the actual content in the text. Instructional activities can include not only previewing text through observing main text features, but moreover through examining and deconstructing the text and ascertaining how the text features can aid them in locating and applying text information for learning.

Included among various text features are a table of contents, headings, subheadings, an index, a glossary, a foreword, a preface, paragraphing spaces, bullet lists, footnotes, sidebars, diagrams, graphs, charts, pictures, illustrations, captions, italics, boldface, colors, and symbols. A glossary is a list of key vocabulary words and/or technical terminology and definitions. This helps readers recognize or learn specialized terms used in the text before reading it. A foreword is typically written by someone other than the text author and appears at the beginning to introduce, inform, recommend, and/or praise the work. A preface is often written by the author and also appears at the beginning, to introduce or explain something about the text, like new additions. A sidebar is a box with text and sometimes graphics at the left or right side of a page, typically focusing on a more specific issue, example, or aspect of the subject. Footnotes are additional comments/notes at the bottom of the page, signaled by superscript numbers in the text.

Text Features on Websites

On the Internet or in computer software programs, text features include URLs, home pages, pop-up menus, drop-down menus, bookmarks, buttons, links, navigation bars, text boxes, arrows, symbols, colors, graphics, logos, and abbreviations. URLs (Universal Resource Locators) indicate the internet "address" or location of a website or web page. They often start with www. (world wide web) or http:// (hypertext transfer protocol) or https:// (the "s" indicates a secure site) and appear in the Internet browser's top address bar. Clickable buttons are often links to specific pages on a website or other external sites. Users can click on some buttons to open pop-up or drop-down menus, which offer a list of actions or departments from which to select. Bookmarks are the electronic versions of physical bookmarks. When users bookmark a website/page, a link is established to the site URL and saved, enabling returning to the site in the future without having to remember its name or URL by clicking the bookmark.

Readers can more easily navigate websites and read their information by observing and utilizing their various text features. For example, most fully developed websites include search bars, where users can type in topics, questions, titles, or names to locate specific information within the large amounts stored on many sites. Navigation bars (software developers frequently use the abbreviation term "navbar") are graphical user interfaces (GUIs) that facilitate visiting different sections, departments, or pages within a website, which can be difficult or impossible to find without these. Typically, they appear as a series of links running horizontally across the top of each page. Navigation bars displayed vertically along the left side of the page are also called sidebars. Links, i.e. hyperlinks, enable hyperspeed browsing by allowing readers to jump to new pages/sites. They may be URLs, words, phrases, images, buttons, etc. They are often but not always underlined and/or blue, or other colors.

Analysis in History/Social Studies and Science

Examining Hypotheses

A **hypothesis** is a well-defined research statement. An experiment then follows, usually using quantitative research. Quantitative research is research based on empirical data.

The results are then analyzed to determine whether the hypothesis was proven or disproven. Examining a hypothesis is also called hypothesis testing. Examining a hypothesis happens most often in science, and it isn't really appropriate for social sciences such as social studies and history. However, qualitative hypotheses can be made in these disciplines to further examine a social or historical event. A hypothesis, in this light, should clearly state the argument that the writer wishes to examine, and the reason or reasons why the author feels it is relevant. This type of hypothesis statement generally requires the "what" and the "why." Consider the following qualitative hypothesis:

"The Métis in Canada were less discriminated against than were Canada's First Nations since they were partly descendants of European fur traders."

The first half of the hypothesis—"The Métis in Canada were less discriminated against than were Canada's First Nations"—reveals the "what," and the second part—"since they were partly descendants of European fur traders"—is the "why."

In science, hypotheses are generally written as "if, then" statements that require the collection of unbiased, empirical, and quantitative data to either prove or disprove the hypothesis. Consider the following:

"**If** a hibiscus flower is placed in direct sunlight and watered twice a day, **then** it will thrive."

The basic steps that lead to the formation of a hypothesis and ultimately, a conclusion, include:

Step 1	Making an observation
Step 2	Forming a question based on the observation
Step 3	Forming a hypothesis (a possible answer to the question)
Step 4	Conducting a study (social studies and history) or an experiment (science)
Step 5	Analyzing the data
Step 6	Drawing a conclusion

In order for conclusions to be accepted as valid and credible, it is extremely important that the data collected isn't biased. The researchers must consider all possible angles of the study or experiment, and they must refrain from collecting the data in such a way as to purposely prove the hypothesis. Conducting studies and experiments of this nature helps to advance the different disciplines, challenge widely accepted beliefs, and broaden a global understanding of the fields of social studies, history, and the sciences.

Interpreting Data

Data can either be **qualitative** (observations, interviews, or focus groups) or **quantitative** (measured data, as in the population of a certain country, a person's height, or the depth of the Earth's oceans). In order for students to interpret and analyze the data, they first must understand the information that has been collected. Collaboration with other students and with professors helps students to further comprehend the collected data. Once fully understood and analyzed, the data can now be interpreted. For instance, how has the analysis of this data affected the students' initial assumptions, thoughts, and beliefs? Data interpretation helps to make a student's knowledge more meaningful. Encouraging metacognitive awareness by asking students to think about how this analysis affects the learning process will also help to strengthen a student's overall understanding and make the learning process more meaningful.

Analysis of data must precede interpretation. Data analysis can take many forms. For instance, students may begin to identify various relationships within the data. Certain patterns or trends may come to light that help students better grasp the meaning or the relevance of the results. The opposite could also be true; students may not identify any patterns, trends, or relationships that would lead to further discussion and evaluation. After fully analyzing the data, students can then begin to interpret this information, which places them in a better position to make informed decisions. Interpreting the data means acquiring a greater understanding of the results of the data. For example, the analysis of collected data on temperature patterns throughout the year and around the globe might reveal that there is a pattern of increasingly hotter temperatures. The interpretation of this data may then result in an individual's greater understanding of global warming.

When students learn to interpret qualitative data in a social studies or historical context, or quantitative data collected in a scientific experiment, their knowledge base expands. The process of data collection, analysis, and interpretation helps to develop an appreciation for how knowledge is attained and why it is critically important to challenge all hypotheses in an attempt to continue to further our collective understanding of the world.

Considering Implications

Any type of research to be conducted should clearly identify the justification for the research as well as outline how the research is to be conducted. Once students have had the opportunity to carry out the research and interpret the results, they must consider the implications. For example, the results might affect the way students will conduct further experiments or research. Data implications could change the way certain industries conduct business, how teachers approach education, or how individuals manage their diet or their health and wellness regime.

The implications of scientific studies may lead medical researchers toward breakthroughs in treatment or guide them that much closer to finding a cure for a disease. In the world of education, data results are the basis for teaching methodologies, and the more data that is collected, analyzed, and interpreted, the more informed the teaching profession is about the optimal ways to prepare for instruction, teaching, and evaluation.

Research on any given topic is multi-faceted. It starts with knowing just enough about a subject to develop a detailed inquiry, which then leads to experimentation or data collection. The data must then be analyzed and interpreted. The final stage is considering what immediate or future implications this discovery may have. Implications of any research study lead to innovative practices across all sectors, provide insights, lead to new discoveries, inspire new research, and create new questions to explore.

Practice Questions

The next five questions are based on the following passage from Martin Luther King Jr.'s "Letter from a Birmingham Jail," 1963.

But more basically, I am in Birmingham because injustice is here. Just as the prophets of the eighth century B.C. left their villages and carried their "thus saith the Lord" far beyond the boundaries of their home towns, and just as the Apostle Paul left his village of Tarsus and carried the gospel of Jesus Christ to the far corners of the Greco-Roman world, so am I compelled to carry the gospel of freedom beyond my own home town. Like Paul, I must constantly respond to the Macedonian call for aid.

Moreover, I am cognizant of the interrelatedness of all communities and states. I cannot sit idly by in Atlanta and not be concerned about what happens in Birmingham. Injustice anywhere is a threat to justice everywhere. We are caught in an inescapable network of mutuality, tied in a single garment of destiny. Whatever affects one directly, affects all indirectly. Never again can we afford to live with the narrow, provincial "outside agitator" idea. Anyone who lives inside the United States can never be considered an outsider anywhere within its bounds ....

In any nonviolent campaign there are four basic steps: 1) collection of the facts to determine whether injustices are alive; 2) negotiation; 3) self-purification; and 4) direct action. We have gone through all of these steps in Birmingham ... Birmingham is probably the most thoroughly segregated city in the United States. Its ugly record of police brutality is known in every section of the country. Its unjust treatment of Negroes in the courts is a notorious reality. There have been more unsolved bombings of Negro homes and churches in Birmingham than in any city in this nation. These are the hard, brutal, and unbelievable facts. On the basis of these conditions Negro leaders sought to negotiate with the city fathers. But the political leaders consistently refused to engage in good faith negotiation.

Then came the opportunity last September to talk with some of the leaders of the economic community. In these negotiating sessions certain promises were made by the merchants—such as the promise to remove the humiliating racial signs from the stores. On the basis of these promises Reverend Shuttlesworth and the leaders of the Alabama Christian Movement for Human Rights agreed to call a moratorium on any type of demonstrations. As the weeks and months unfolded we realized that we were the victims of a broken promise. The signs remained. As in so many experiences in the past, we were confronted with blasted hopes, and the dark shadow of a deep disappointment settled upon us. So we had no alternative except that of preparing for direct action, whereby we would present our very bodies as a means of laying our case before the conscience of the local and national community. We were not unmindful of the difficulties involved. So we decided to go through the process of self-purification. We started having workshops on nonviolence and repeatedly asked ourselves the questions, "Are you able to accept the blows without retaliating?" "Are you able to endure the ordeals of jail?"

1. What is the main idea of this passage?
   a. Birmingham was one of the most segregated cities in the United States.
   b. Injustice anywhere is a threat to justice everywhere.
   c. Direct action is necessary to get results.
   d. One cannot trust civic leaders to negotiate.

2. In the passage, what does the word "moratorium" mean:
   a. Temporary stoppage
   b. Temporary increase
   c. Continuation
   d. Resumption

3. In the sentence: "We are caught in an inescapable network of mutuality, tied in a single garment of destiny," *garment* is a:
   a. Simile
   b. Allusion
   c. Metaphor
   d. Symbolism

4. Based on the passage, what left King and other Black leaders no choice except to act directly?
   a. Bombings continued
   b. Humiliating racial signs remained in windows
   c. Merchants refused to meet with them
   d. There was no time to gather facts and information

5. That King repeatedly brings up religion is MOST likely related to which of the following?
   a. He knew that religion would be a strong ethical appeal to individuals reading the text.
   b. He was a reverend and would therefore use scripture.
   c. He worked with the Alabama Christian Movement and knew they would appreciate his inclusion of scripture.
   d. He was writing to religious leaders who would respond best to the inclusion of religious references.

Answer Explanations

1. B: King comes right out and says, "injustice anywhere is a threat to justice everywhere," thereby clarifying his message and explaining why he has traveled to Birmingham. While Choices *A*, *C*, and *D* are discussed in the passage, they are not the main focus. Choice *A* is an example of why King feels the need to travel to Birmingham. Choice *C* is what he argues we must do to respond. And, finally Choice *D* is a problem he runs into specifically when attempting to negotiate rather than act.

2. A: A moratorium is a temporary stoppage, in this case of the protests, so business leaders could address concerns.

3. C: A metaphor is a comparison of two things that are not alike and is not literally possible. In other words, people are not literally in a garment, but their futures are likely woven together like clothing.

4. B: While other problems might have continued in the Birmingham community, the only one that King mentions specifically in the passage is "The signs remained." Therefore, Choice *B* is the correct answer.

5. A: While Choices *B*, *C*, and *D* are all true, King's audience was much wider than those in the religious community and his role was more than that of a religious leader. Instead, he chooses this tactic to appeal to Southerners for whom religion was an important aspect of their lives. He can, in this way, appeal to their values and make connections between those values and his goals.

SAT Writing and Language Test

Standard English Conventions

Conventions of Standard English Spelling

Homophones

Homophones are words that have different meanings and spellings, but sound the same. These can be confusing for English Language Learners (ELLs) and beginning students, but even native English-speaking adults can find them problematic unless informed by context. Whereas listeners must rely entirely on context to differentiate spoken homophone meanings, readers with good spelling knowledge have a distinct advantage since homophones are spelled differently. For instance, *their* means belonging to them; *there* indicates location; and *they're* is a contraction of *they are*; despite different meanings, they all sound the same. *Lacks* can be a plural noun or a present-tense, third-person singular verb; either way it refers to absence—*deficiencies* as a plural noun, and *is deficient in* as a verb. But *lax* is an adjective that means loose, slack, relaxed, uncontrolled, or negligent. These two spellings, derivations, and meanings are completely different. With speech, listeners cannot know spelling and must use context; but with print, readers with spelling knowledge can differentiate them with or without context.

Homonyms, Homophones, and Homographs

Homophones are words that sound the same in speech but have different spellings and meanings. For example, *to, too,* and *two* all sound alike, but have three different spellings and meanings. Homophones with different spellings are also called **heterographs. Homographs** are words that are spelled identically but have different meanings. If they also have different pronunciations, they are **heteronyms.** For instance, *tear* pronounced one way means a drop of liquid formed by the eye; pronounced another way, it means to rip. Homophones that are also homographs are **homonyms**. For example, *bark* can mean the outside of a tree or a dog's vocalization; both meanings have the same spelling. *Stalk* can mean a plant stem or to pursue and/or harass somebody; these are spelled and pronounced the same. *Rose* can mean a flower or the past tense of *rise.* Many non-linguists confuse things by using "homonym" to mean sets of words that are homophones but not homographs, and also those that are homographs but not homophones.

The word *row* can mean to use oars to propel a boat; a linear arrangement of objects or print; or an argument. It is pronounced the same with the first two meanings, but differently with the third. Because it is spelled identically regardless, all three meanings are homographs. However, the two meanings pronounced the same are homophones, whereas the one with the different pronunciation is a heteronym. By contrast, the word *read* means to peruse language, whereas the word *reed* refers to a marsh plant. Because these are pronounced the same way, they are homophones; because they are spelled differently, they are heterographs. Homonyms are both homophones and homographs—pronounced and spelled identically, but with different meanings. One distinction between homonyms is of those with separate, unrelated etymologies, called "true" homonyms, e.g. *skate* meaning a fish or *skate* meaning to glide over ice/water. Those with common origins are called **polysemes** or **polysemous homonyms**, e.g. the *mouth* of an animal/human or of a river.

Irregular Plurals

While many words in English can become plural by adding –s or –es to the end, there are some words that have irregular plural forms. One type includes words that are spelled the same whether they are singular or plural, such as deer, fish, salmon, trout, sheep, moose, offspring, species, aircraft, etc. The spelling rule for making these words plural is simple: they do not change. Other irregular English plurals change form based on vowel shifts, linguistic mutations, or grammatical and spelling conventions from their languages of origin, like Latin or German. Some examples include *child* and *children;* *die* and *dice;* *foot* and *feet;* *goose* and *geese;* *louse* and *lice;* *man* and *men;* *mouse* and *mice;* *ox* and *oxen;* *person* and *people;* *tooth* and *teeth;* and *woman* and *women.*

Contractions

Contractions are formed by joining two words together, omitting one or more letters from one of the component words, and replacing the omitted letter(s) with an apostrophe. An obvious yet often forgotten rule for spelling contractions is to place the apostrophe where the letters were omitted. For example, didn't is a contraction of did not; therefore, the apostrophe replaces the "o" that is omitted from the "not." Another common error is confusing contractions with possessives because both include apostrophes, e.g. spelling the possessive *its* as "it's," which is a contraction of "it is"; spelling the possessive *their* as "they're," a contraction of "they are"; spelling the possessive *whose* as "who's," a contraction of "who is"; or spelling the possessive *your* as "you're," a contraction of "you are."

Frequently Misspelled Words

One source of spelling errors is not knowing whether to drop the final letter *e* from a word when its form is changed; some words retain the final *e* when another syllable is added while others lose it. For example, *true* becomes *truly;* *argue* becomes *arguing;* *come* becomes *coming;* *write* becomes *writing;* and *judge* becomes *judging.* In these examples, the final *e* is dropped before adding the ending. But *severe* becomes *severely;* *complete* becomes *completely;* *sincere* becomes *sincerely;* *argue* becomes *argued;* and *care* becomes *careful.* In these instances, the final *e* is retained before adding the ending. Note that some words, like argue in these examples, drops the final e when the –ing ending is added to indicate the participial form, but the regular past tense form keeps the e and adds a –d to make it argued.

Other commonly misspelled English words are those containing the vowel combinations ei and ie. Many people confuse these two. Some examples of words with the ei combination include:

> *ceiling, conceive, leisure, receive, weird, their, either, foreign, sovereign, neither, neighbors, seize, forfeit, counterfeit, height, weight, protein,* and *freight*

Words with *ie* include *piece, believe, chief, field, friend, grief, relief, mischief, siege, niece, priest, fierce, pierce, achieve, retrieve, hygiene, science,* and *diesel.* A rule that also functions as a mnemonic device is "I before E except after C, or when sounded like A as in 'neighbor' or 'weigh'." However, it is obvious from the list above that many exceptions exist.

People often misspell certain words by confusing whether they have the vowel a, e, or i. For example, in the following correctly spelled words, the vowel in boldface is the one people typically get wrong by substituting one of the others for it:

> *cem**e**tery, quant**i**ties, ben**e**fit, priv**i**lege, unpleas**a**nt, sep**a**rate, independ**e**nt, excell**e**nt, cat**e**gories, indispens**a**ble,* and *irrelev**a**nt*

Some words with final syllables that sound the same when spoken but are spelled differently include *unpleasant, independent, excellent,* and *irrelevant.* Another source of misspelling is whether or not to double consonants when adding suffixes. For example, double the last consonant before –ed and –ing endings in controlled, beginning, forgetting, admitted, occurred, referred, and hopping; but do not double before the suffix in *shining, poured, sweating, loving, hating, smiling,* and *hoping.*

One final example of common misspellings involves either the failure to include silent letters or the converse of adding extraneous letters. If a letter is not pronounced in speech, it is easy to leave it out in writing. For example, some people omit the silent *u* in *g**u**arantee,* overlook the first *r* in *su**r**prise,* leave out the *z* in *reali**z**e,* fail to double the *m* in *reco**mm**end,* leave out the middle *i* from *asp**i**rin,* and exclude the *p* from *tem**p**erature.* The converse error, adding extra letters, is common in words like *until* by adding a second *l* at the end; or by inserting a superfluous syllabic *a* or *e* in the middle of *athletic,* reproducing a common mispronunciation.

Conventions of Standard English Punctuation

Rules of Capitalization

The first word of any document, and of each new sentence, is capitalized. Proper nouns, like names and adjectives derived from proper nouns, should also be capitalized. Here are some examples:

- Grand Canyon
- Pacific Palisades
- Golden Gate Bridge
- Freudian slip
- Shakespearian, Spenserian, or Petrarchan sonnet
- Irish song

Some exceptions are adjectives, originally derived from proper nouns, which through time and usage are no longer capitalized, like *quixotic, herculean,* or *draconian.* Capitals draw attention to specific instances of people, places, and things. Some categories that should be capitalized include the following:

- brand names
- companies
- weekdays
- months
- governmental divisions or agencies
- historical eras
- major historical events
- holidays
- institutions
- famous buildings
- ships and other manmade constructions
- natural and manmade landmarks
- territories
- nicknames
- epithets
- organizations
- planets
- nationalities
- tribes
- religions
- names of religious deities

- roads
- special occasions, like the Cannes Film Festival or the Olympic Games

Exceptions
Related to American government, capitalize the noun Congress but not the related adjective congressional. Capitalize the noun U.S. Constitution, but not the related adjective constitutional. Many experts advise leaving the adjectives federal and state in lowercase, as in federal regulations or state water board, and only capitalizing these when they are parts of official titles or names, like Federal Communications Commission or State Water Resources Control Board. While the names of the other planets in the solar system are capitalized as names, Earth is more often capitalized only when being described specifically as a planet, like Earth's orbit, but lowercase otherwise since it is used not only as a proper noun but also to mean *land, ground, soil,* etc. While the name of the Bible should be capitalized, the adjective biblical should not. Regarding biblical subjects, the words heaven, hell, devil, and satanic should not be capitalized. Although race names like Caucasian, African-American, Navajo, Eskimo, East Indian, etc. are capitalized, white and black as races are not.

Names of animal species or breeds are not capitalized unless they include a proper noun. Then, only the proper noun is capitalized. Antelope, black bear, and yellow-bellied sapsucker are not capitalized. However, Bengal tiger, German shepherd, Australian shepherd, French poodle, and Russian blue cat are capitalized.

Other than planets, celestial bodies like the sun, moon, and stars are not capitalized. Medical conditions like tuberculosis or diabetes are lowercase; again, exceptions are proper nouns, like Epstein-Barr syndrome, Alzheimer's disease, and Down syndrome. Seasons and related terms like winter solstice or autumnal equinox are lowercase. Plants, including fruits and vegetables, like poinsettia, celery, or avocados, are not capitalized unless they include proper names, like Douglas fir, Jerusalem artichoke, Damson plums, or Golden Delicious apples.

Titles and Names
When official titles precede names, they should be capitalized, except when there is a comma between the title and name. But if a title follows or replaces a name, it should not be capitalized. For example, "the president" without a name is not capitalized, as in "The president addressed Congress." But with a name it is capitalized, like "President Obama addressed Congress." Or, "Chair of the Board Janet Yellen was appointed by President Obama." One exception is that some publishers and writers nevertheless capitalize President, Queen, Pope, etc., when these are not accompanied by names to show respect for these high offices. However, many writers in America object to this practice for violating democratic principles of equality. Occupations before full names are not capitalized, like owner Mark Cuban, director Martin Scorsese, or coach Roger McDowell.

Some universal rules for capitalization in composition titles include capitalizing the following:

- The first and last words of the title
- Forms of the verb *to be* and all other verbs
- Pronouns
- The word *not*

Universal rules for NOT capitalizing include the articles *the, a,* or *an;* the conjunctions *and, or,* or *nor,* and the preposition *to* or *to* as part of the infinitive form of a verb. The exception to all of these is UNLESS any of them is the first or last word in the title, in which case they are capitalized. Other words are subject to differences of opinion and differences among various stylebooks or methods. These include *as, but, if,* and *or,* which some capitalize and others do not. Some authorities say no preposition should ever be capitalized; some say prepositions five or more letters long should be capitalized. The *Associated Press Stylebook* advises capitalizing prepositions longer than three letters (like *about, across,* or *with*).

Ellipses

Ellipses (. . .) signal omitted text when quoting. Some writers also use them to show a thought trailing off, but this should not be overused outside of dialogue. An example of an ellipsis would be if someone is quoting a phrase out of a professional source but wants to omit part of the phrase that isn't needed: "Dr. Skim's analysis of pollen inside the body is clearly a myth . . . that speaks to the environmental guilt of our society."

Commas

Commas separate words or phrases in a series of three or more. The **Oxford comma** is the last comma in a series. Many people omit this last comma, but many times it causes confusion. Here is an example:

> I love my sisters, the Queen of England and Madonna.

This example without the comma implies that the "Queen of England and Madonna" are the speaker's sisters. However, if the speaker was trying to say that they love their sisters, the Queen of England, as well as Madonna, there should be a comma after "Queen of England" to signify this.

Commas also separate two coordinate adjectives ("big, heavy dog") but not cumulative ones, which should be arranged in a particular order for them to make sense ("beautiful ancient ruins").

A comma ends the first of two independent clauses connected by conjunctions. Here is an example:

> I ate a bowl of tomato soup, and I was hungry very shortly after.

Here are some brief rules for commas:

- Commas follow introductory words like *however, furthermore, well, why,* and *actually,* among others.
- Commas go between city and state: Houston, Texas.
- If using a comma between a surname and Jr. or Sr. or a degree like M.D., also follow the whole name with a comma: "Martin Luther King, Jr., wrote that."
- A comma follows a dependent clause beginning a sentence: "Although she was very small, . . ."
- Nonessential modifying words/phrases/clauses are enclosed by commas: "Wendy, who is Peter's sister, closed the window."
- Commas introduce or interrupt direct quotations: "She said, 'I hate him.' 'Why,' I asked, 'do you hate him?'"

Semicolons

Semicolons are used to connect two independent clauses but should never be used in the place of a comma. They can replace periods between two closely connected sentences: "Call back tomorrow; it can wait until then." When writing items in a series and one or more of them contains internal commas, separate them with semicolons, like the following:

> People came from Springfield, Illinois; Alamo, Tennessee; Moscow, Idaho; and other locations.

Hyphens

Here are some rules concerning hyphens:

- Compound adjectives like state-of-the-art or off-campus are hyphenated.
- Original compound verbs and nouns are often hyphenated, like "throne-sat," "video-gamed," "no-meater."
- Adjectives ending in *–ly* are often hyphenated, like "family-owned" or "friendly-looking."
- "Five years old" is not hyphenated, but singular ages like "five-year-old" are.
- Hyphens can clarify. For example, in "stolen vehicle report," "stolen-vehicle report" clarifies that "stolen" modifies "vehicle," not "report."
- Compound numbers twenty-one through ninety-nine are spelled with hyphens.
- Prefixes before proper nouns/adjectives are hyphenated, like "mid-September" and "trans-Pacific."

Parentheses

Parentheses enclose information such as an aside or more clarifying information: "She ultimately replied (after deliberating for an hour) that she was undecided." They are also used to insert short, in-text definitions or acronyms: "His FBS (fasting blood sugar) was higher than normal." When parenthetical information ends the sentence, the period follows the parentheses: "We received new funds ($25,000)." Only put periods within parentheses if the whole sentence is inside them: "Look at this. (You'll be astonished.)" However, this can also be acceptable as a clause: "Look at this (you'll be astonished)." Although parentheses appear to be part of the sentence subject, they are not, and do not change subject-verb agreement: "Will (and his dog) was there."

Quotation Marks

Quotation marks are typically used when someone is quoting a direct word or phrase someone else writes or says. Additionally, quotation marks should be used for the titles of poems, short stories, songs, articles, chapters, and other shorter works. When quotations include punctuation, periods and commas should *always* be placed inside of the quotation marks.

When a quotation contains another quotation inside of it, the outer quotation should be enclosed in double quotation marks and the inner quotation should be enclosed in single quotation marks. For example: "Timmy was begging, 'Don't go! Don't leave!'" When using both double and single quotation marks, writers will find that many word-processing programs may automatically insert enough space between the single and double quotation marks to be visible for clearer reading. But if this is not the case, the writer should write/type them with enough space between to keep them from looking like three single quotation marks. Additionally, non-standard usages, terms used in an unusual fashion, and technical terms are often clarified by quotation marks. Here are some examples:

> My "friend," Dr. Sims, has been micromanaging me again.
>
> This way of extracting oil has been dubbed "fracking."

Apostrophes

One use of the apostrophe is followed by an *s* to indicate possession, like *Mrs. White's home* or *our neighbor's dog*. When using the *'s* after names or nouns that also end in the letter *s*, no single rule applies: some experts advise adding both the apostrophe and the *s*, like "the Jones's house," while others prefer using only the apostrophe and omitting the additional *s*, like "the Jones' house." The wisest expert advice is to pick one formula or the other and then apply it consistently. Newspapers and magazines often use *'s* after common nouns ending with *s*, but add only the apostrophe after proper nouns or names ending with *s*. One common error is to place the apostrophe before a name's final *s* instead of after it: "Ms. Hasting's book" is incorrect if the name is Ms. Hastings.

Plural nouns should not include apostrophes (e.g. "apostrophe's"). Exceptions are to clarify atypical plurals, like verbs used as nouns: "These are the do's and don'ts." Irregular plurals that do not end in *s* always take apostrophe-*s*, not *s*-apostrophe—a common error, as in "childrens' toys," which should be "children's toys." Compound nouns like mother-in-law, when they are singular and possessive, are followed by apostrophe-*s*, like "your mother-in-law's coat." When a compound noun is plural and possessive, the plural is formed before the apostrophe-*s*, like "your sisters-in-laws' coats." When two people named possess the same thing, use apostrophe-*s* after the second name only, like "Dennis and Pam's house."

Sentence Structures

Incomplete Sentences
Four types of incomplete sentences are sentence fragments, run-on sentences, subject-verb and/or pronoun-antecedent disagreement, and non-parallel structure.

Sentence fragments are caused by absent subjects, absent verbs, or dangling/uncompleted dependent clauses. Every sentence must have a subject and a verb to be complete. An example of a fragment is "Raining all night long," because there is no subject present. "It was raining all night long" is one correction. Another example of a sentence fragment is the second part in "Many scientists think in unusual ways. Einstein, for instance." The second phrase is a fragment because it has no verb. One correction is "Many scientists, like Einstein, think in unusual ways." Finally, look for "cliffhanger" words like *if, when, because,* or *although* that introduce dependent clauses, which cannot stand alone without an independent clause. For example, to correct the sentence fragment "If you get home early," add an independent clause: "If you get home early, we can go dancing."

Run-On Sentences
A **run-on sentence** combines two or more complete sentences without punctuating them correctly or separating them. For example, a run-on sentence caused by a lack of punctuation is the following:

> There is a malfunction in the computer system however there is nobody available right now who knows how to troubleshoot it.

One correction is, "There is a malfunction in the computer system; however, there is nobody available right now who knows how to troubleshoot it." Another is, "There is a malfunction in the computer system. However, there is nobody available right now who knows how to troubleshoot it."

An example of a comma splice of two sentences is the following:

> Jim decided not to take the bus, he walked home.

Replacing the comma with a period or a semicolon corrects this. Commas that try and separate two independent clauses without a contraction are considered comma splices.

Parallel Sentence Structures
Parallel structure in a sentence matches the forms of sentence components. Any sentence containing more than one description or phrase should keep them consistent in wording and form. Readers can easily follow writers' ideas when they are written in parallel structure, making it an important element of correct sentence construction. For example, this sentence lacks parallelism: "Our coach is a skilled manager, a clever strategist, and works hard." The first two phrases are parallel, but the third is not. Correction: "Our coach is a skilled manager, a clever strategist, and a hard worker." Now all three phrases match in form.

Here is another example:

> Fred intercepted the ball, escaped tacklers, and a touchdown was scored.

This is also non-parallel. Here is the sentence corrected:

> Fred intercepted the ball, escaped tacklers, and scored a touchdown.

Sentence Fluency

For fluent composition, writers must use a variety of sentence types and structures, and also ensure that they smoothly flow together when they are read. To accomplish this, they must first be able to identify fluent writing when they read it. This includes being able to distinguish among simple, compound, complex, and compound-complex sentences in text; to observe variations among sentence types, lengths, and beginnings; and to notice figurative language and understand how it augments sentence length and imparts musicality. Once students/writers recognize superior fluency, they should revise their own writing to be more readable and fluent. They must be able to apply acquired skills to revisions before being able to apply them to new drafts.

One strategy for revising writing to increase its sentence fluency is flipping sentences. This involves rearranging the word order in a sentence without deleting, changing, or adding any words. For example, the student or other writer who has written the sentence, "We went bicycling on Saturday" can revise it to, "On Saturday, we went bicycling." Another technique is using appositives. An appositive is a phrase or word that renames or identifies another adjacent word or phrase. Writers can revise for sentence fluency by inserting main phrases/words from one shorter sentence into another shorter sentence, combining them into one longer sentence, e.g. from "My cat Peanut is a gray and brown tabby. He loves hunting rats." to "My cat Peanut, a gray and brown tabby, loves hunting rats." Revisions can also connect shorter sentences by using conjunctions and commas and removing repeated words: "Scott likes eggs. Scott is allergic to eggs" becomes "Scott likes eggs, but he is allergic to them."

One technique for revising writing to increase sentence fluency is "padding" short, simple sentences by adding phrases that provide more details specifying why, how, when, and/or where something took place. For example, a writer might have these two simple sentences: "I went to the market. I purchased a cake." To revise these, the writer can add the following informative dependent and independent clauses and prepositional phrases, respectively: "Before my mother woke up, I sneaked out of the house and went to the supermarket. As a birthday surprise, I purchased a cake for her." When revising sentences to make them longer, writers must also punctuate them correctly to change them from simple sentences to compound, complex, or compound-complex sentences.

Skills Writers Can Employ to Increase Fluency

One way writers can increase fluency is by varying the beginnings of sentences. Writers do this by starting most of their sentences with different words and phrases rather than monotonously repeating the same ones across multiple sentences. Another way writers can increase fluency is by varying the lengths of sentences. Since run-on sentences are incorrect, writers make sentences longer by also converting them from simple to compound, complex, and compound-complex sentences. The coordination and subordination involved in these also give the text more variation and interest, hence more fluency.

Here are a few more ways writers can increase fluency:

- Varying the transitional language and conjunctions used makes sentences more fluent.
- Writing sentences with a variety of rhythms by using prepositional phrases.
- Varying sentence structure adds fluency.

Expression of Ideas

Elements of the Writing Process

Skilled writers undergo a series of steps that comprise the writing process. The purpose of adhering to a structured approach to writing is to develop clear, meaningful, coherent work.

The stages are pre-writing or planning, organizing, drafting/writing, revising, and editing. Not every writer will necessarily follow all five stages for every project, but will judiciously employ the crucial components of the stages for most formal or important work. For example, a brief informal response to a short reading passage may not necessitate the need for significant organization after idea generation, but larger assignments and essays will likely mandate use of the full process.

Pre-Writing/Planning

Brainstorming

One of the most important steps in writing is pre-writing. Before drafting an essay or other assignment, it's helpful to think about the topic for a moment or two, in order to gain a more solid understanding of what the task is. Then, spend about five minutes jotting down the immediate ideas that could work for the essay. Brainstorming is a way to get some words on the page and offer a reference for ideas when drafting. Scratch paper is provided for writers to use any pre-writing techniques such as webbing, freewriting, or listing. Some writers prefer using graphic organizers during this phase. The goal is to get ideas out of the mind and onto the page.

Freewriting

Like brainstorming, freewriting is another prewriting activity to help the writer generate ideas. This method involves setting a timer for two or three minutes and writing down all ideas that come to mind about the topic using complete sentences. Once time is up, writers should review the sentences to see what observations have been made and how these ideas might translate into a more unified direction for the topic. Even if sentences lack sense as a whole, freewriting is an excellent way to get ideas onto the page in the very beginning stages of writing. Using complete sentences can make this a bit more challenging than brainstorming, but overall it is a worthwhile exercise, as it may force the writer to come up with more complete thoughts about the topic.

Once the ideas are on the page, it's time for the writer to turn them into a solid plan for the essay. The best ideas from the brainstorming results can then be developed into a more formal outline.

Organizing

Although sometimes it is difficult to get going on the brainstorming or prewriting phase, once ideas start flowing, writers often find that they have amassed too many thoughts that will not make for a cohesive and unified essay. During the organization stage, writers should examine the generated ideas, hone in on the important ones central to their main idea, and arrange the points in a logical and effective manner. Writers may also determine that some of the ideas generated in the planning process need further elaboration, potentially necessitating the need for research to gather information to fill the gaps.

Once a writer has chosen his or her thesis and main argument, selected the most applicable details and evidence, and eliminated the "clutter," it is time to strategically organize the ideas. This is often accomplished with an outline.

Outlining

An **outline** is a system used to organize writing. When composing essays, outlining is important because it helps writers organize important information in a logical pattern using Roman numerals. Usually, outlines start out with the main ideas and then branch out into subgroups or subsidiary thoughts or subjects. Not only do outlines provide a visual tool for writers to reflect on how events, ideas, evidence, or other key parts of the argument relate to one another, but they can also lead writers to a stronger conclusion. The sample below demonstrates what a general outline looks like:

I. Introduction
  1. Background
  2. Thesis statement

II. Body
  1. Point A
    a. Supporting evidence
    b. Supporting evidence
  2. Point B
    a. Supporting evidence
    b. Supporting evidence
  3. Point C
    a. Supporting evidence
    b. Supporting evidence

III. Conclusion
  1. Restatement of main points.
  2. Memorable ending.

Drafting/Writing

Now it comes time to actually write the essay. In this stage, writers should follow the outline they developed in the brainstorming process and try to incorporate the useful sentences penned in the freewriting exercise. The main goal of this phase is to put all the thoughts together in cohesive sentences and paragraphs.

It is helpful for writers to remember that their work here does not have to be perfect. This process is often referred to as **drafting** because writers are just creating a rough draft of their work. Because of this, writers should avoid getting bogged down on the small details.

Referencing Sources

Anytime a writer quotes or paraphrases another text, they will need to include a citation. A citation is a short description of the work that a quote or information came from. The style manual your teacher wants you to follow will dictate exactly how to format that citation. For example, this is how one would cite a book according to the APA manual of style:

- *Format*: Last name, First initial, Middle initial. (Year Published) *Book Title*. City, State: Publisher.
- *Example*: Sampson, M. R. (1989). *Diaries from an alien invasion*. Springfield, IL: Campbell Press.

Revising

Revising offers an opportunity for writers to polish things up. Putting one's self in the reader's shoes and focusing on what the essay actually says helps writers identify problems—it's a movement from the mindset of writer to the mindset of editor. The goal is to have a clean, clear copy of the essay.

The main goal of the revision phase is to improve the essay's flow, cohesiveness, readability, and focus. For example, an essay will make a less persuasive argument if the various pieces of evidence are scattered and presented illogically or clouded with unnecessary thought. Therefore, writers should consider their essay's structure and organization, ensuring that there are smooth transitions between sentences and paragraphs. There should be a discernable introduction and conclusion as well, as these crucial components of an essay provide readers with a blueprint to follow.

Additionally, if the writer includes copious details that do little to enhance the argument, they may actually distract readers from focusing on the main ideas and detract from the strength of their work. The ultimate goal is to retain the purpose or focus of the essay and provide a reader-friendly experience. Because of this, writers often need to delete parts of their essay to improve its flow and focus. Removing sentences, entire paragraphs, or large chunks of writing can be one of the toughest parts of the writing process because it is difficult to part with work one has done. However, ultimately, these types of cuts can significantly improve one's essay.

Lastly, writers should consider their voice and word choice. The voice should be consistent throughout and maintain a balance between an authoritative and warm style, to both inform and engage readers. One way to alter voice is through word choice. Writers should consider changing weak verbs to stronger ones and selecting more precise language in areas where wording is vague. In some cases, it is useful to modify sentence beginnings or to combine or split up sentences to provide a more varied sentence structure.

Editing

Rather than focusing on content (as is the aim in the revising stage), the editing phase is all about the mechanics of the essay: the syntax, word choice, and grammar. This can be considered the proofreading stage. Successful editing is what sets apart a messy essay from a polished document.

The following areas should be considered when proofreading:

- Sentence fragments
- Awkward sentence structure
- Run-on sentences
- Incorrect word choice
- Grammatical agreement errors
- Spelling errors
- Punctuation errors
- Capitalization errors

One of the most effective ways of identifying grammatical errors, awkward phrases, or unclear sentences is to read the essay out loud. Listening to one's own work can help move the writer from simply the author to the reader.

During the editing phase, it's also important to ensure the essay follows the correct formatting and citation rules as dictated by the assignment.

Recursive Writing Process

While the writing process may have specific steps, the good news is that the process is recursive, meaning the steps need not be completed in a particular order. Many writers find that they complete steps at the same time such as drafting and revising, where the writing and rearranging of ideas occur simultaneously or in very close order. Similarly, a writer may find that a particular section of a draft needs more development and will go back to the prewriting stage to generate new ideas. The steps can be repeated at any time, and the more these steps of the recursive writing process are employed, the better the final product will be.

Practice Makes Prepared Writers

Like any other useful skill, writing only improves with practice. While writing may come more easily to some than others, it is still a skill to be honed and improved. Regardless of a person's natural abilities, there is always room for growth in writing. Practicing the basic skills of writing can aid in preparations for the SAT.

One way to build vocabulary and enhance exposure to the written word is through reading. This can be through reading books, but reading of any materials such as newspapers, magazines, and even social media count towards practice with the written word. This also helps to enhance critical reading and thinking skills, through analysis of the ideas and concepts read. Think of each new reading experience as a chance to sharpen these skills.

Practice Questions

The next five questions are based on the following passage.

The morning sun shone down upon the land and the sky was a deep blue. The snowcapped mountains in the distance seemed a thousand miles away, and the valleys down below were covered in the most beautiful yellow flowers. It was springtime (1) once again in the Arctic, and (2) just like countless times before, signs of life were beginning to reappear. The polar bear, foxes, seal, and hare's (3) would soon appear, along with countless newborns to carry on the lifeline.

Throughout the winter months, it seems as though no life exists at all in the Arctic. The winter winds are chillingly cold, and the water is frozen solid. The only sounds for months on end are the distant cries of a lone wolf (4) or the howling wind ravaging through the land. Everything is white and crisp, but some startling changes are beginning to show. The sea ice is melting at an alarming rate, and for the people who call the Arctic their home, this is devastating news. Melting ice means it will be more and more difficult to hunt. Hunters use the frozen fjord as their highway to hunt for fresh food. As the frozen land disappears, so too is their way of life. (5) The people of the Arctic have lived off of the land and have an immense respect for nature, an attitude that has passed been down through the generations. They are worried that the rest of the world has become detached, separated from the Earth. It is because of this detachment that the Earth now suffers. In years past, in springtime there would still be enough frozen land to allow hunters to travel to the best hunting grounds, but now there is so much water that the hunters cannot travel the necessary distance. Their entire villages rely on their skills, but without enough frozen land, hunters are returning home with very little food. The one message the people of the Arctic would like to pass on to the rest of the world is very simple—no matter what you are doing in this moment, take the time to get reconnected to nature. There is still time.

1. Choose the best replacement.
   a. NO CHANGE
   b. spring time
   c. Springtime
   d. spring-time

2. Choose the best replacement.
   a. NO CHANGE
   b. arctic and
   c. arctic, and
   d. arctic, and,

3. Choose the best replacement.
   a. NO CHANGE
   b. rabbits
   c. hare
   d. hair

4. Choose the best replacement.
    a. NO CHANGE
    b. loan wolf
    c. Lone Wolf
    d. lone-wolf

5. Choose the best replacement sentence.
    a. NO CHANGE
    b. As the frozen land disappears, so is their way of life.
    c. As the frozen land disappears, so too, is their way of life.
    d. As the frozen land disappears, so too does their way of life.

Answer Explanations

1. A: The best answer is Choice *A*. No change is needed here, because "springtime" is one compound word. It is not two words, as in Choice *B*. It does not require a hyphen, as in Choice *C*. It is not capitalized, as in Choice *D*.

2. A: Choice *A* is the best answer. No change is needed here, because "Arctic," when used as a noun, is capitalized.

3. C: The best answer is Choice *C*. A hare is a long-eared animal similar to a rabbit. The word "hare" can be plural; it is also correct to use "hares" as the plural. Choice *A* is not the best answer because it is in the possessive form. Choice *B* is not the best answer because rabbits are slightly different than hares; for example, most rabbits live underground, while hares nest above ground.

4. A: The best answer is Choice *A*. No change is needed. Choice *B* is incorrect, because the word "loan" refers to something borrowed. Choice *C* is incorrect, because there is no reason to capitalize the phrase "lone wolf." No hyphen is needed, so Choice *D* is incorrect.

5. D: Choice *D* is the best answer. The verb "so does" has to follow the phrase that comes before it—that the land is disappearing. Let's put it in a sentence: "their way of life does disappear" or "their way of life is disappear"? With the latter we would have to change the verb to "disappearing."

SAT Math Test

Heart of Algebra

Creating, Solving, or Interpreting a Linear Expression or Equation in One Variable

An **equation in one variable** is a mathematical statement where two algebraic expressions in one variable, usually x, are set equal. To solve the equation, the variable must be isolated on one side of the equals sign. The addition and multiplication principles of equality are used to isolate the variable. The **addition principle of equality** states that the same number can be added to or subtracted from both sides of an equation. Because the same value is being used on both sides of the equals sign, equality is maintained.

For example, the equation $2x = 5x$ is equivalent to both:

$$2x + 3 = 5x + 3$$

and

$$2x - 5 = 5x - 5$$

This principle can be used to solve the following equation: $x + 5 = 4$. The variable x must be isolated, so to move the 5 from the left side, subtract 5 from both sides of the equals sign.

Therefore:

$$x + 5 - 5 = 4 - 5$$

So, the solution is $x = -1$. This process illustrates the idea of an **additive inverse** because subtracting 5 is the same as adding -5. Basically, add the opposite of the number that must be removed to both sides of the equals sign. The **multiplication principle of equality** states that equality is maintained when a number is either multiplied times both expressions on each side of the equals sign, or when both expressions are divided by the same number.

For example, $4x = 5$ is equivalent to both $16x = 20$ and $x = \frac{5}{4}$. Multiplying both sides times 4 and dividing both sides by 4 maintains equality. Solving the equation:

$$6x - 18 = 5$$

requires the use of both principles. First, apply the addition principle to add 18 to both sides of the equals sign, which results in $6x = 23$. Then use the multiplication principle to divide both sides by 6, giving the solution $x = \frac{23}{6}$. Using the multiplication principle in the solving process is the same as involving a multiplicative inverse. A **multiplicative inverse** is a value that, when multiplied by a given number, results in 1. Dividing by 6 is the same as multiplying by $\frac{1}{6}$, which is both the reciprocal and multiplicative inverse of 6.

When solving a linear equation in one variable, checking the answer shows if the solution process was performed correctly. Plug the solution into the variable in the original equation. If the result is a false statement, something was done incorrectly during the solution procedure. Checking the example above gives the following:

$$6 \times \frac{23}{6} - 18 = 23 - 18 = 5$$

Therefore, the solution is correct.

Some equations in one variable involve fractions or the use of the distributive property. In either case, the goal is to obtain only one variable term and then use the addition and multiplication principles to isolate that variable.

Consider the equation $\frac{2}{3}x = 6$. To solve for x, multiply each side of the equation by the reciprocal of $\frac{2}{3}$, which is $\frac{3}{2}$. This step results in:

$$\frac{3}{2} \times \frac{2}{3}x = \frac{3}{2} \times 6$$

which simplifies into the solution $x = 9$. Now consider the equation:

$$3(x + 2) - 5x = 4x + 1$$

Use the distributive property to clear the parentheses. Therefore, multiply each term inside the parentheses by 3. This step results in:

$$3x + 6 - 5x = 4x + 1$$

Next, collect like terms on the left-hand side. **Like terms** are terms with the same variable or variables raised to the same exponent(s). Only like terms can be combined through addition or subtraction. After collecting like terms, the equation is:

$$-2x + 6 = 4x + 1$$

Finally, apply the addition and multiplication principles. Add $2x$ to both sides to obtain $6 = 6x + 1$. Then, subtract 1 from both sides to obtain $5 = 6x$. Finally, divide both sides by 6 to obtain the solution $\frac{5}{6} = x$.

Two other types of solutions can be obtained when solving an equation in one variable. The final result could be that there is either no solution or that the solution set contains all real numbers. Consider the equation:

$$4x = 6x + 5 - 2x$$

First, the like terms can be combined on the right to obtain:

$$4x = 4x + 5$$

Next, subtract $4x$ from both sides. This step results in the false statement $0 = 5$. There is no value that can be plugged into x that will ever make this equation true. Therefore, there is no solution. The solution procedure contained correct steps, but the result of a false statement means that no value satisfies the equation. The symbolic way to denote that no solution exists is $\emptyset$. Next, consider the equation:

$$5x + 4 + 2x = 9 + 7x - 5$$

Combining the like terms on both sides results in:

$$7x + 4 = 7x + 4$$

The left-hand side is exactly the same as the right-hand side. Using the addition principle to move terms, the result is $0 = 0$, which is always true. Therefore, the original equation is true for any number, and the solution set is all real numbers. The symbolic way to denote such a solution set is $\mathbb{R}$, or in interval notation, $(-\infty, \infty)$.

Translating Phrases and Sentences into Expressions, Equations, and Inequalities

When presented with a real-world problem, the first step is to determine what unknown quantity must be solved for. Use a variable, such as x or t, to represent that unknown quantity. Sometimes there can be two or more unknown quantities. In this case, either choose an additional variable, or if a relationship exists between the unknown quantities, express the other quantities in terms of the original variable. After choosing the variables, form algebraic expressions and/or equations that represent the verbal statement in the problem. The following table shows examples of vocabulary used to represent the different operations.

Addition	Sum, plus, total, increase, more than, combined, in all
Subtraction	Difference, less than, subtract, reduce, decrease, fewer, remain
Multiplication	Product, multiply, times, part of, twice, triple
Division	Quotient, divide, split, each, equal parts, per, average, shared

The combination of operations and variables form both mathematical expression and equations. The differences between expressions and equations are that there is no equals sign in an expression, and that expressions are evaluated to find an unknown quantity, while equations are solved to find an unknown quantity. Also, inequalities can exist within verbal mathematical statements. Instead of a statement of equality, expressions state quantities are *less than*, *less than or equal to*, *greater than*, or *greater than or equal to*. Another type of inequality is when a quantity is said to be not equal to another quantity ($\neq$).

The steps for solving inequalities in one variable are the same steps for solving equations in one variable. The addition and multiplication principles are used. However, to maintain a true statement when using the $<, \leq, >$, and $\geq$ symbols, if a negative number is either multiplied times both sides of an inequality or divided from both sides of an inequality, the sign must be flipped. For instance, consider the following inequality:

$$3 - 5x \leq 8$$

First, 3 is subtracted from each side to obtain $-5x \leq 5$. Then, both sides are divided by -5, while flipping the sign, to obtain $x \geq -1$. Therefore, any real number greater than or equal to -1 satisfies the original inequality.

Solving Real-World One- or Multi-Step Problems with Rational Numbers

One-step problems take only one mathematical step to solve. For example, solving the equation $5x = 45$ is a one-step problem because the one step of dividing both sides of the equation by 5 is the only step necessary to obtain the solution $x = 9$. The **multiplication principle of equality** is the one step used to isolate the variable. The equation is of the form $ax = b$, where a and b are rational numbers. Similarly, the **addition principle of equality** could be the one step needed to solve a problem. In this case, the equation would be of the form $x + a = b$ or $x - a = b$, for real numbers a and b.

A **multi-step problem** involves more than one step to find the solution, or it could consist of solving more than one equation. An equation that involves both the addition principle and the multiplication principle is a two-step problem, and an example of such an equation is $2x - 4 = 5$. To solve, add 4 to both sides and then divide both sides by 2. An example of a two-step problem involving two separate equations is $y = 3x, 2x + y = 4$. The two equations form a system that must be solved together in two variables. The system can be solved by the substitution method. Since y is already solved for in terms of x, replace y with $3x$ in the equation $2x + y = 4$, resulting in $2x + 3x = 4$. Therefore, $5x = 4$ and $x = \frac{4}{5}$. Because there are two variables, the solution consists of a value for both x and for y. Substitute $x = \frac{4}{5}$ into either original equation to find y. The easiest choice is $y = 3x$. Therefore:

$$y = 3 \times \frac{4}{5} = \frac{12}{5}$$

The solution can be written as the ordered pair $\left(\frac{4}{5}, \frac{12}{5}\right)$.

Real-world problems can be translated into both one-step and multi-step problems. In either case, the word problem must be translated from the verbal form into mathematical expressions and equations that can be solved using algebra. An example of a one-step real-world problem is the following: A cat weighs half as much as a dog living in the same house.

If the dog weighs 14.5 pounds, how much does the cat weigh? To solve this problem, an equation can be used. In any word problem, the first step must be defining variables that represent the unknown quantities. For this problem, let x be equal to the unknown weight of the cat. Because two times the weight of the cat equals 14.5 pounds, the equation to be solved is: $2x = 14.5$. Use the multiplication principle to divide both sides by 2. Therefore, $x = 7.25$, and the cat weighs 7.25 pounds.

Most of the time, real-world problems require multiple steps. The following is an example of a multi-step problem: The sum of two consecutive page numbers is equal to 437. What are those page numbers? First, define the unknown quantities. If x is equal to the first page number, then $x + 1$ is equal to the next page number because they are consecutive integers. Their sum is equal to 437. Putting this information together results in the equation:

$$x + x + 1 = 437$$

To solve, first collect like terms to obtain $2x + 1 = 437$.

Then, subtract 1 from both sides and then divide by 2. The solution to the equation is $x = 218$. Therefore, the two consecutive page numbers that satisfy the problem are 218 and 219. It is always important to make sure that answers to real-world problems make sense. For instance, it should be a red flag if the solution to this same problem resulted in decimals, which would indicate the need to check the work. Page numbers are whole numbers; therefore, if decimals are found to be answers, the solution process should be double-checked for mistakes.

Creating, Solving, or Interpreting a Linear Inequality in One Variable

A **linear equation** in *x* can be written in the form $ax + b = 0$. A **linear inequality** is very similar, although the equals sign is replaced by an inequality symbol such as $<, >, \leq,$ or $\geq$. In any case, *a* can never be 0. Some examples of linear inequalities in one variable are:

$$2x + 3 < 0$$

and

$$4x - 2 \leq 0$$

Solving an inequality involves finding the set of numbers that when plugged into the variable, make the inequality a true statement. These numbers are known as the **solution set** of the inequality. To solve an inequality, use the same properties that are necessary in solving equations. First, add or subtract variable terms and/or constants to obtain all variable terms on one side of the equals sign and all constant terms on the other side. Then, either multiply both sides times the same number, or divide both sides by the same number, to obtain an inequality that gives the solution set. When multiplying times, or dividing by, a negative number in an inequality, change the direction of the inequality symbol. The solution set can be graphed on a number line. Consider the linear inequality:

$$-2x - 5 > x + 6$$

First, add 5 to both sides and subtract $-x$ off of both sides to obtain $-3x > 11$. Then, divide both sides by -3, making sure to change the direction of the inequality symbol. These steps result in the solution:

$$x < -\frac{11}{3}$$

Therefore, any number less than $-\frac{11}{3}$ satisfies this inequality.

Building a Linear Function that Models a Linear Relationship Between Two Quantities

A **linear function that models a linear relationship between two quantities** is of the form $y = mx + b$ or in function form:

$$f(x) = mx + b$$

In a linear function, the value of *y* depends on the value of *x*, and *y* increases or decreases at a constant rate as *x* increases. Therefore, the independent variable is *x*, and the dependent variable is *y*. The graph of a linear function is a line, and the constant rate can be seen by looking at the steepness, or slope, of the line. If the line increases from left to right, the slope is positive. If the line slopes downward from left to right, the slope is negative. In the function, *m* represents slope. Each point on the line is an **ordered pair** (*x*, *y*), where *x* represents the *x*-coordinate of the point and *y* represents the *y*-coordinate of the point. The point where *x* = 0 is known as the *y*-intercept, and it is the place where the line crosses the *y*-axis. If *x* = 0 is plugged into $f(x) = mx + b$, the result is $f(0) = b$, so therefore, the point (0, *b*) is the *y*-intercept of the line. The derivative of a linear function is its slope.

Consider the following situation. A taxicab driver charges a flat fee of $2 per ride and $3 a mile. This statement can be modeled by the function $f(x) = 3x + 2$ where x represents the number of miles and $f(x) = y$ represents the total cost of the ride. The total cost increases at a constant rate of $2 per mile, and that is why this situation is a linear relationship. The slope $m = 3$ is equivalent to this rate of change.

The flat fee of $2 is the y-intercept. It is the place where the graph crosses the x-axis, and it represents the cost when $x = 0$, or when no miles have been traveled in the cab. The y-intercept in this situation represents the flat fee.

Creating, Solving, or Interpreting Systems of Linear Inequalities in Two Variables

A **system of linear inequalities in two variables** consists of two inequalities in two variables, x and y. For example, the following is a system of linear inequalities in two variables:

$$\begin{cases} 4x + 2y < 1 \\ 2x - y \leq 0 \end{cases}$$

The curly brace on the left side shows that the two inequalities are grouped together. A solution of a single inequality in two variables is an ordered pair that satisfies the inequality. For example, (1, 3) is a solution of the linear inequality $y \geq x + 1$ because when plugged in, it results in a true statement. The graph of an inequality in two variables consists of all ordered pairs that make the solution true. Therefore, the entire solution set of a single inequality contains many ordered pairs, and the set can be graphed by using a half plane. A **half plane** consists of the set of all points on one side of a line. If the inequality consists of $>$ or $<$, the line is dashed because no solutions actually exist on the line shown. If the inequality consists of $\geq$ or $\leq$, the line is solid and solutions are on the line shown. To graph a linear inequality, graph the corresponding equation found by replacing the inequality symbol with an equals sign. Then pick a test point that exists on either side of the line. If that point results in a true statement when plugged into the original inequality, shade in the side containing the test point. If it results in a false statement, shade in the opposite side.

Solving a system of linear inequalities must be done graphically. Follow the process as described above for both given inequalities. The solution set to the entire system is the region that is in common to every graph in the system.

For example, here is the solution to the following system:

$$\begin{cases} y \geq 3 - x \\ y \leq -3 - x \end{cases}$$

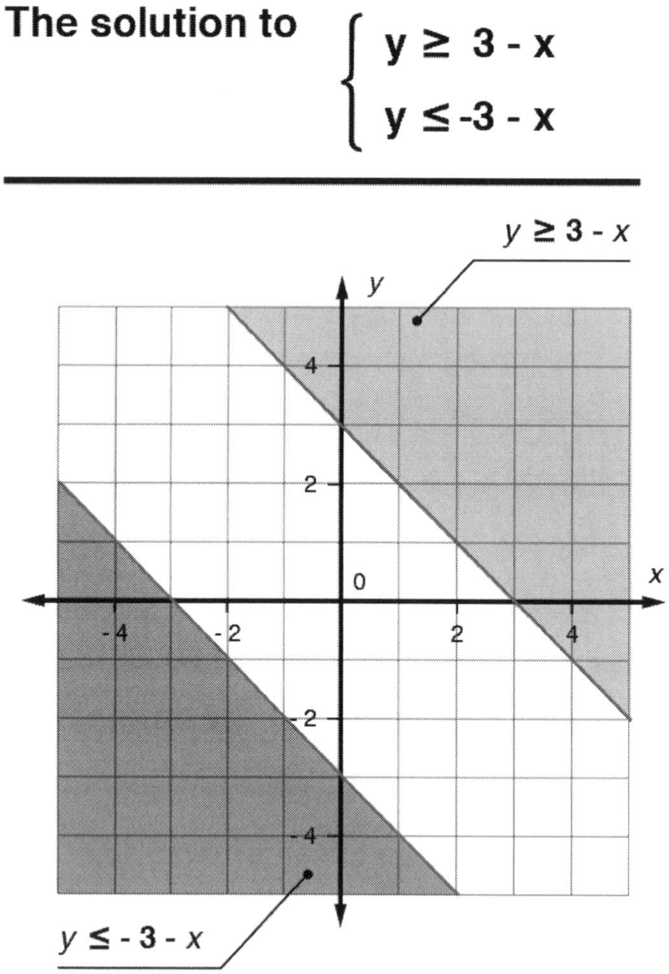

Note that there is no region in common, so this system has no solution.

Creating, Solving, or Interpreting Systems of Two Linear Equations in Two Variables

An example of a **system of two linear equations in two variables** is the following:

$$2x + 5y = 8$$

$$5x + 48y = 9$$

A solution to a system of two linear equations is an ordered pair that satisfies both the equations in the system. A system can have one solution, no solution, or infinitely many solutions. The solution can be found through a graphing technique. The solution of a system of equations is actually equal to the point of intersection of both lines. If the lines intersect at one point, there is one solution and the system is said to be **consistent**. However, if the two lines are parallel, they will never intersect and there is no solution. In this case, the system is said to be **inconsistent.** Third, if the two lines are actually the same line, there are infinitely many solutions and the solution set is equal to the entire line. The lines are **dependent**. Here is a summary of the three cases:

Solving Systems by Graphing

Consistent	**Inconsistent**	**Dependent**
One solution	No solution	Infinite number of solutions
Lines intersect	*Lines are parallel*	*Coincide: same line*

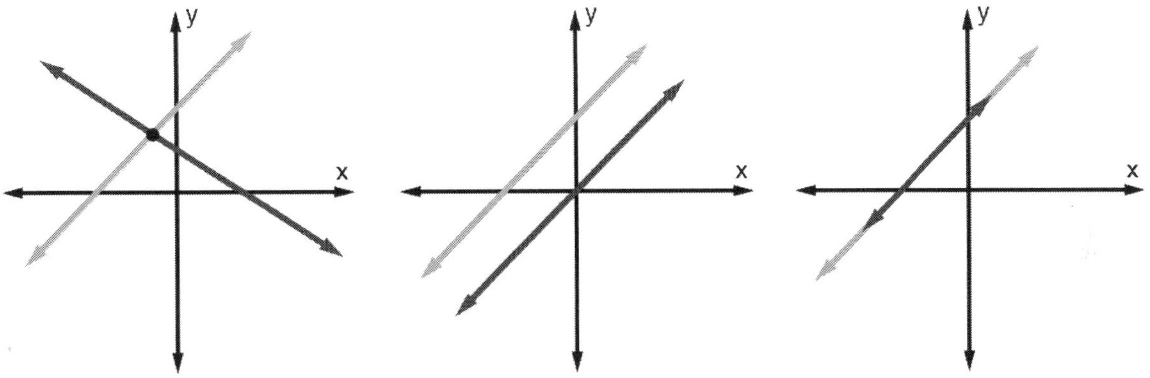

Consider the following system of equations:

$$y + x = 3$$

$$y - x = 1$$

To find the solution graphically, graph both lines on the same *xy*-plane. Graph each line using either a table of ordered pairs, the *x*- and *y*-intercepts, or slope and the *y*-intercept. Then, locate the point of intersection. The graph is shown here:

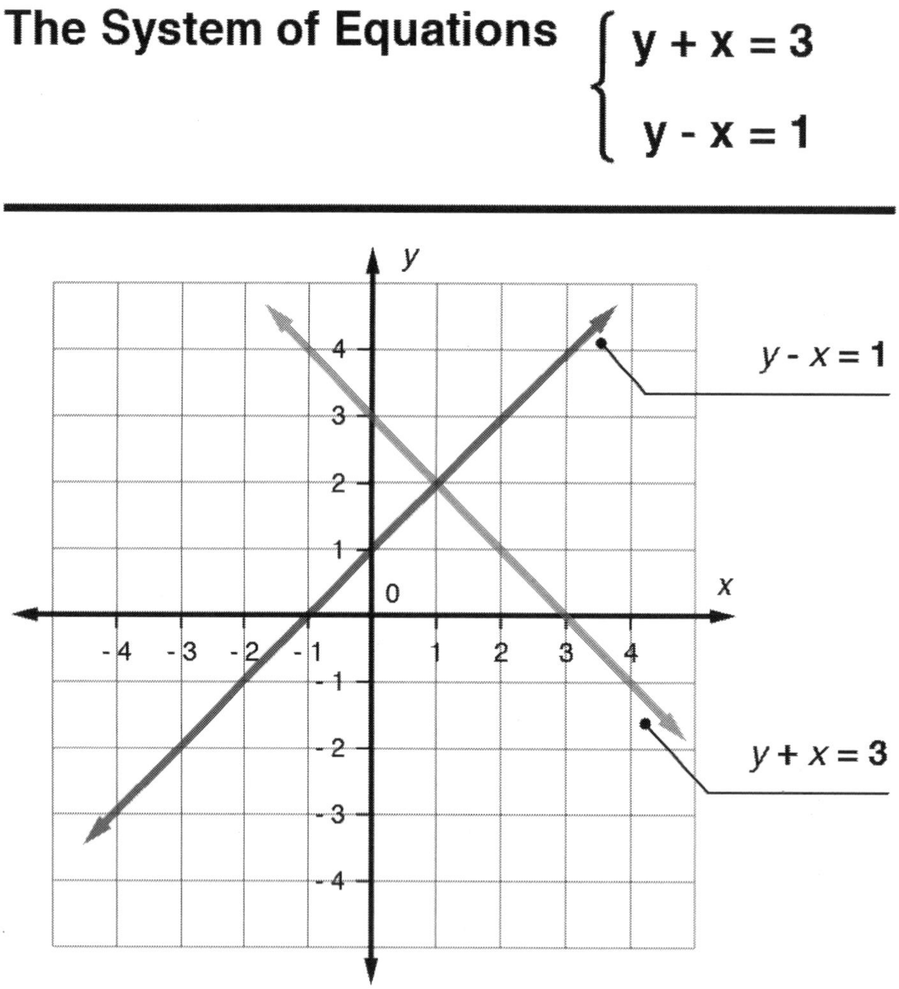

It can be seen that the point of intersection is the ordered pair (1, 2). This solution can be checked by plugging it back into both original equations to make sure it results in true statements. This process results in:

$$2 + 1 = 3$$

$$2 - 1 = 1$$

Both are true equations, so therefore the point of intersection is truly the solution.

The following system has no solution:

$$y = 4x + 1$$

$$y = 4x - 1$$

Both lines have the same slope and different y-intercepts, so they are parallel, meaning that they run alongside each other and never intersect.

Finally, the following solution has infinitely many solutions:

$$2x - 7y = 12$$

$$4x - 14y = 24$$

Note that the second equation is equal to the first equation times 2. Therefore, they are the same line. The solution set can be written in set notation as $\{(x, y)|2x - 7y = 12\}$, which represents the entire line.

Algebraically Solving Linear Equations or Inequalities in One Variable

A **linear equation in one variable** can be solved using the following steps:

1. Simplify the algebraic expressions on both sides of the equals sign by removing all parentheses, using the distributive property, and then collect all like terms.

2. Collect all variable terms on one side of the equals sign and all constant terms on the other side by adding the same quantity to both sides of the equals sign, or by subtracting the same quantity from both sides of the equals sign.

3. Isolate the variable by either dividing both sides of the equation by the same number, or by multiplying both sides by the same number.

4. Check the answer.

The only difference between solving linear inequalities versus equations is that when multiplying by a negative number or dividing by a negative number, the direction of the inequality symbol must be reversed.

If an equation contains multiple fractions, it might make sense to clear the equation of fractions first by multiplying all terms by the least common denominator. Also, if an equation contains several decimals, it might make sense to clear the decimals as well by multiplying times a factor of 10. If the equation has decimals in the hundredths place, multiply every term in the equation by 100.

Algebraically Solving Systems of Two Linear Equations in Two Variables

There are two algebraic methods to finding solutions. The first is substitution. This process is better suited for systems when one of the equations is already solved for one variable, or when solving for one variable is easy to do. The equation that is already solved for is substituted into the other equation for that variable, and this process results in a linear equation in one variable. This equation can be solved for the given variable, and then that solution can be plugged into one of the original equations, which can then be solved for the other variable. This last step is known as **back-substitution**, and the end result is an ordered pair.

The following is an example of a system that is suited for substitution:

$$y = 4x + 2$$

$$2x + 3y = 9$$

The other method is known as **elimination**, or the **addition method**. This is better suited when the equations are in standard form $Ax + By = C$. The goal in this method is to multiply one or both equations times numbers that result in opposite coefficients. Then, add the equations together to obtain an equation in one variable. Solve for the given variable, then take that value and back-substitute to obtain the other part of the ordered pair solution.

The following is an example of a system that is suited for elimination:

$$2x + 3y = 8$$

$$4x - 2y = 10$$

Note that in order to check an answer when solving a system of equations, the solution must be checked in both original equations to show that it solves not only one of the equations, but both of them.

If either solution results in an untrue statement when inserted into the original equation, then there is no solution to the system. Finally, if throughout either solution procedure the process results in the variables dropping out, which gives a statement that is always true, there are infinitely many solutions.

Interpreting the Variables and Constants in Expressions for Linear Functions within the Context Presented

A **linear function** of the form $f(x) = mx + b$ has two important quantities: m and b. The quantity m represents the slope of the line, and the quantity b represents the y-intercept of the line. When the function represents a real-life situation or a mathematical model, these two quantities are very meaningful. The slope, m, represents the rate of change, or the amount y increases or decreases given an increase in x. If m is positive, the rate of change is positive, and if m is negative, the rate of change is negative. The y-intercept, b, represents the amount of quantity y when x is 0. In many applications, if the x-variable is never a negative quantity, the y-intercept represents the initial amount of the quantity y. The x-variable often represents time, so it makes sense that it would not be negative.

Consider the following example. These two equations represent the cost, C, of t-shirts, x, at two different printing companies:

$$C(x) = 7x$$

$$C(x) = 5x + 25$$

The first equation represents a scenario in which each t-shirt costs $7. In this equation, x varies directly with y. There is no y-intercept, which means that there is no initial cost for using that printing company. The rate of change is 7, which is price per shirt. The second equation represents a scenario that has both an initial cost and a cost per t-shirt. The slope of 5 shows that each shirt is $5. The y-intercept of 25 shows that there is an initial cost of using that company. Therefore, it makes sense to use the first company at $7 per shirt when only purchasing a small number of t-shirts. However, any large orders would be cheaper from the second company because eventually that initial cost would become negligible.

Understanding Connections Between Algebraic and Graphical Representations

They all organize, categorize, and compare data, and they come in different shapes and sizes. Each type has its own way of showing information, whether through a column, shape, or picture. To answer a question relating to a table, chart, or graph, some steps should be followed. First, the problem should be read thoroughly to determine what is being asked to determine what quantity is unknown. Then, the title of the table, chart, or graph should be read. The title should clarify what data is actually being summarized in the table. Next, look at the key and labels for both the horizontal and vertical axes, if they are given. These items will provide information about how the data is organized. Finally, look to see if there is any more labeling inside the table. Taking the time to get a good idea of what the table is summarizing will be helpful as it is used to interpret information.

Tables are a good way of showing a lot of information in a small space. The information in a table is organized in columns and rows. For example, a table may be used to show the number of votes each candidate received in an election. By interpreting the table, one may observe which candidate won the election and which candidates came in second and third. In using a bar chart to display monthly rainfall amounts in different countries, rainfall can be compared between countries at different times of the year. Graphs are also a useful way to show change in variables over time, as in a line graph, or percentages of a whole, as in a pie graph.

The table below relates the number of items to the total cost. The table shows that one item costs $5. By looking at the table further, five items cost $25, ten items cost $50, and fifty items cost $250. This cost can be extended for any number of items. Since one item costs $5, then two items would cost $10. Though this information is not in the table, the given price can be used to calculate unknown information.

Number of Items	1	5	10	50
Cost ($)	5	25	50	250

A **bar graph** is a graph that summarizes data using bars of different heights. It is useful when comparing two or more items or when seeing how a quantity changes over time. It has both a horizontal and vertical axis. To interpret bar graphs, recognize what each bar represents and connect that to the two variables. The bar graph below shows the scores for 6 people on 3 different games. The different colors of the bars distinguish between the three games, and the height of the bar indicates their score for that game. James scored 42 on game 1, and Robert scored 50 on game 2. By comparing the bars, it's obvious that Emily scored higher than William. By observing the exact scores, Abigail scored 14 points higher on game 2 than Olivia.

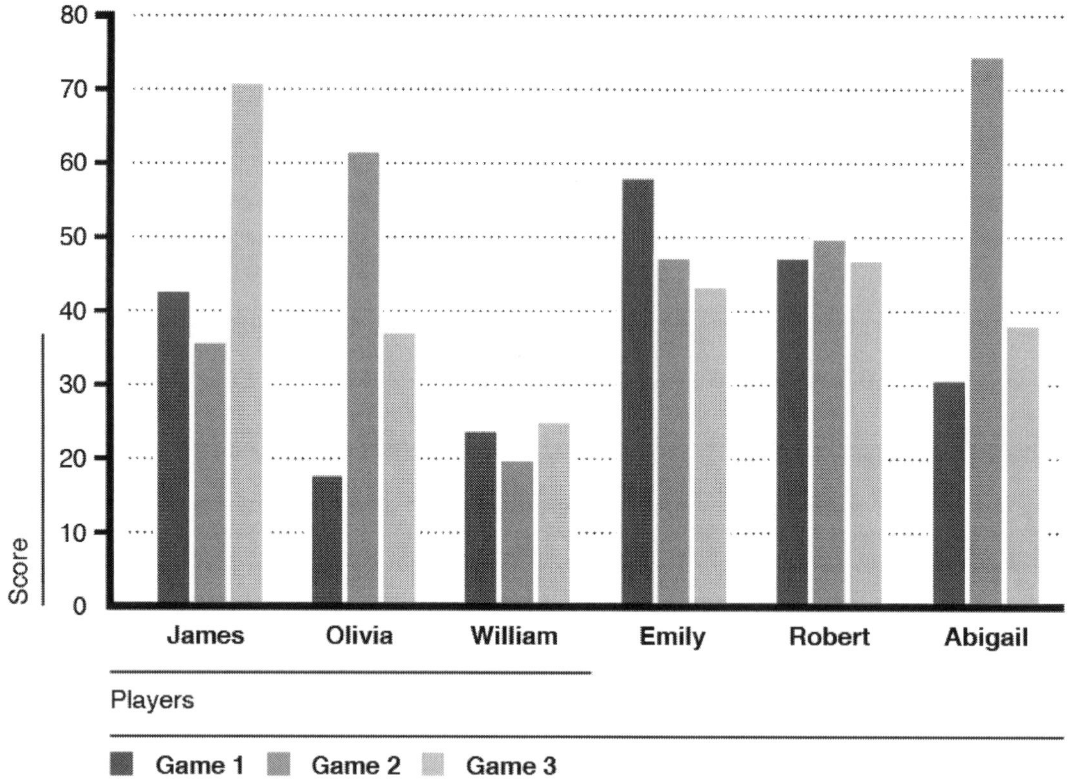

A line graph is a way to compare two variables that are plotted on opposite axes of a graph. The line indicates a continuous change as it rises or falls. The line's rate of change is known as its slope. The horizontal axis often represents a variable of time. Audiences can quickly see if an amount has grown or decreased over time. The bottom of the graph, or the x-axis, shows the units for time, such as days, hours, months, etc. If there are multiple lines, a comparison can be made between what the two lines represent. For example, the following line graph shows the change in temperature over five days. The line with the triangles represents the high, and the line with the squares represents the low for each day. Looking at the triangle line alone, the high decreases at an almost constant rate for two days, then decreases more on Thursday, and rises again on Friday. The low temperatures have a similar trend, shown in the square line.

The range in temperatures each day can also be calculated by finding the difference between the square line and the triangle line on a particular day. On Wednesday, the range was 10 degrees, from 65 to 75°F.

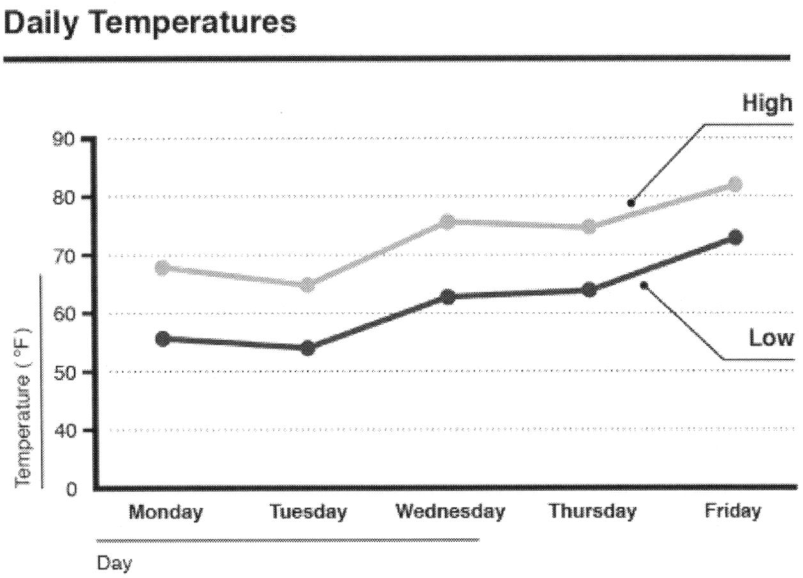

Pie charts are used to show percentages of a whole, as each category is given a piece of the pie, and together all the pieces make up a whole. They are a circular representation of data which are used to highlight numerical proportion. It is true that the arc length of each pie slice is proportional to the amount it individually represents. When a pie chart is shown, an audience can quickly make comparisons by comparing the sizes of the pieces of the pie. They can be useful for comparison between different categories. The following pie chart is a simple example of three different categories shown in comparison to each other. From the labels, black represents dogs, dark gray represents other pets, and light gray represents cats. These three equal pieces each represent just more than 33 percent, or $\frac{1}{3}$ of the whole. Dogs and cats may be combined to represent $\frac{2}{3}$ of the whole.

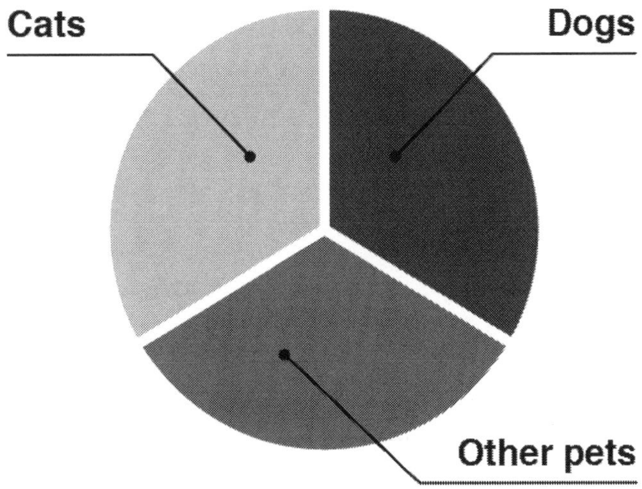

The fact that a circle is 360 degrees is used to create a pie chart. Because each piece of the pie is a percentage of a whole, that percentage is multiplied times 360 to get the number of degrees each piece represents. In the example above, each piece is $\frac{1}{3}$ of the whole, so each piece is equivalent to 120 degrees. Together, all three pieces add up to 360 degrees.

Stacked bar graphs, also used fairly frequently, are used when comparing multiple variables at one time. They combine some elements of both pie charts and bar graphs, using the organization of bar graphs and the proportionality aspect of pie charts. The following is an example of a stacked bar graph that represents the number of students in a band playing drums, flute, trombone, and clarinet. Each bar graph is broken up further into girls and boys.

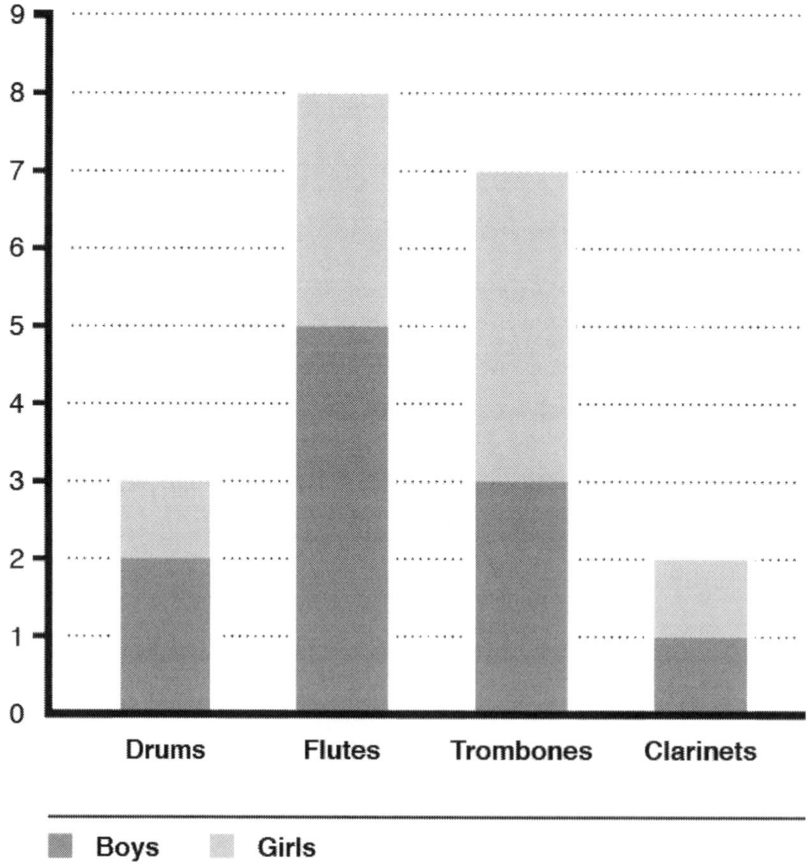

To determine how many boys play trombone, refer to the darker portion of the trumpet bar, resulting in 3 students.

A **scatterplot** is another way to represent paired data. It uses Cartesian coordinates, like a line graph, meaning it has both a horizontal and vertical axis. Each data point is represented as a dot on the graph.

The dots are never connected with a line. For example, the following is a scatterplot showing the connection between people's ages and heights.

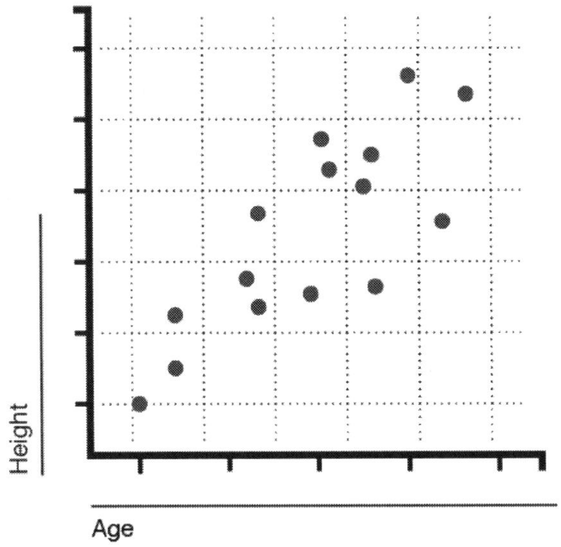

A scatterplot, also known as a **scattergram**, can be used to predict another value and to see if an association, known as a **correlation**, exists between a set of data. If the data resembles a straight line, the data is **associated.** The following is an example of a scatterplot in which the data does not seem to have an association:

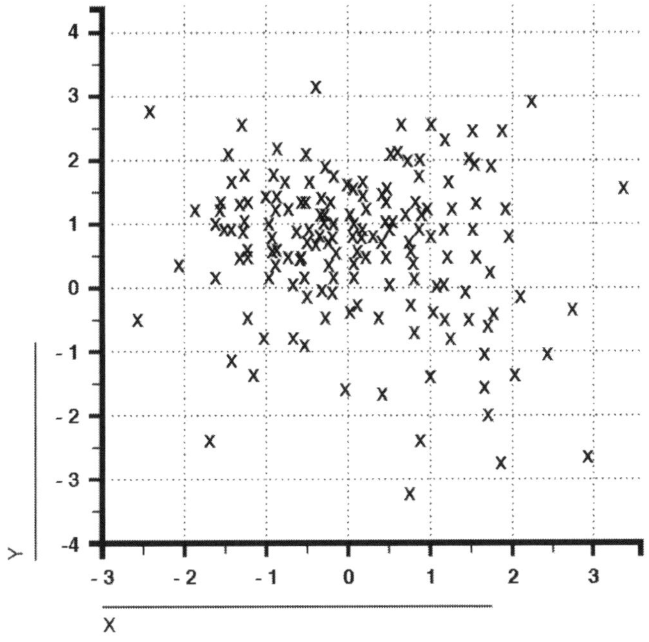

Sets of numbers and other similarly organized data can also be represented graphically. Venn diagrams are a common way to do so. A **Venn diagram** represents each set of data as a circle. The circles overlap, showing that each set of data is overlapping. A Venn diagram is also known as a **logic diagram** because it visualizes all possible logical combinations between two sets. Common elements of two sets are represented by the area of overlap. The following is an example of a Venn diagram of two sets A and B:

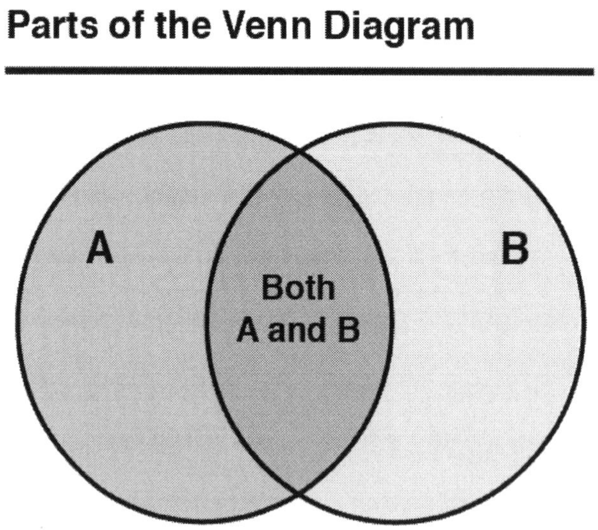

Another name for the area of overlap is the **intersection.** The intersection of A and B, $A \cap B$, contains all elements that are in both sets A and B. The **union** of A and B, $A \cup B$, contains all elements in both sets A and B. Finally, the **complement** of $A \cup B$ is equal to all elements that are not in either set A or set B. These elements are placed outside of the circles.

The following is an example of a Venn diagram in which 30 students were surveyed asking which type of siblings they had: brothers, sisters, or both. Ten students only had a brother, seven students only had a sister, and five had both a brother and a sister. Therefore, five is the intersection, represented by the section where the circles overlap. Two students did not have a brother or a sister. Two is therefore the complement and is placed outside of the circles.

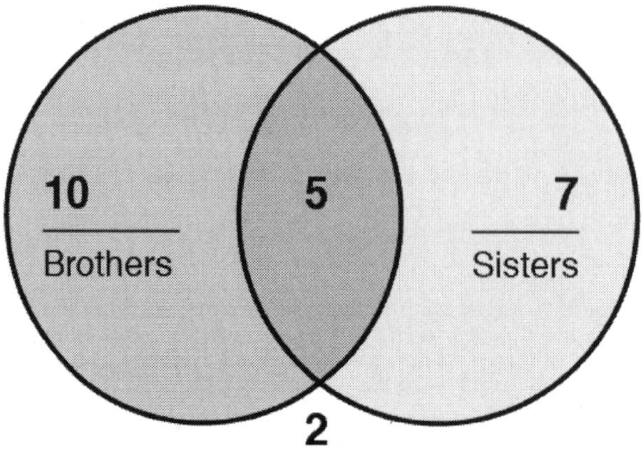

Venn diagrams can have more than two sets of data. The more circles, the more logical combinations are represented by the overlapping. The following is a Venn diagram that represents a different situation. Now, there were 30 students surveyed about the color of their socks. The innermost region represents those students that have green, pink, and blue socks on (perhaps a striped pattern). Therefore, two students had all three colors. on their socks. In this example, all students had at least one of the three colors on their socks, so no one exists in the complement.

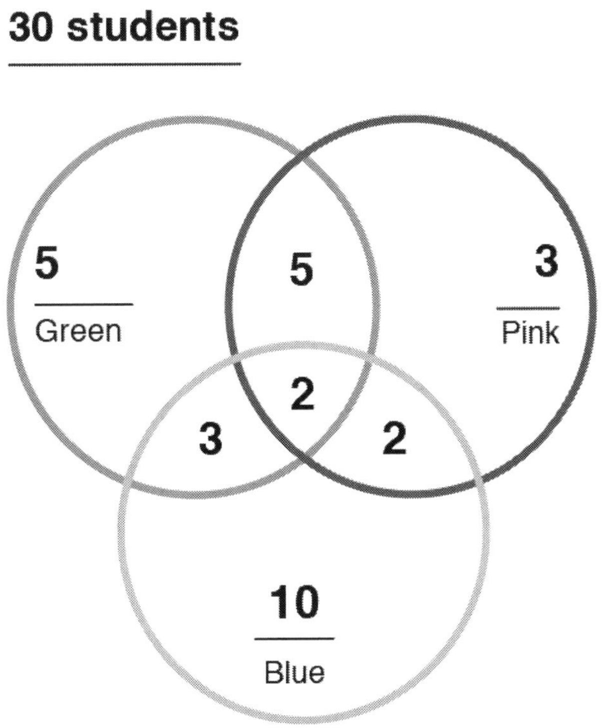

Venn diagrams are typically not drawn to scale; however, if they are, and if each circle's area is proportional to the amount of data it represents, then it is called an area-proportional Venn diagram.

Problem Solving and Data Analysis

Using Ratios, Rates, Proportional Relationships, and Scale Drawings to Solve Single- and Multistep Problems

Solving Real-World Problems Involving Ratios and Rates of Change

Recall that a **ratio** is the comparison of two different quantities. Comparing 2 apples to 3 oranges results in the ratio 2:3, which can be expressed as the fraction $\frac{2}{5}$. Note that order is important when discussing ratios. The number mentioned first is the antecedent, and the number mentioned second is the consequent. Note that the consequent of the ratio and the denominator of the fraction are *not* the same. When there are 2 apples to 3 oranges, there are five fruit total; two fifths of the fruit are apples, while three fifths are oranges. The ratio 2:3 represents a different relationship that the ratio 3:2. Also, it is important to make sure that when discussing ratios that have units attached to them, the two quantities use the same units. For example, to compare 8 feet to 4 yards, Therefore, the ratio would be 8 feet to 12 feet, which can be expressed as the fraction $\frac{8}{20}$. Also, note that it is proper to refer to ratios in lowest terms. Therefore, the ratio of 8 feet to 4 yards is equivalent to the fraction $\frac{2}{5}$.

Many real-world problems involve ratios. Often, problems with ratios involve proportions, as when two ratios are set equal to find the missing amount. However, some problems involve deciphering single ratios. For example, consider an amusement park that sold 345 tickets last Saturday. If 145 tickets were sold to adults and the rest of the tickets were sold to children, what would the ratio of the number of adult tickets to children's tickets be? A common mistake would be to say the ratio is 145:345. However, 345 is the total number of tickets sold, not the number of children's tickets. There were $345 - 145 = 200$ tickets sold to children. The correct ratio of adult to children's tickets is 145:200. As a fraction, this expression is written as $\frac{145}{345}$, which can be reduced to $\frac{29}{69}$.

While a ratio compares two measurements using the same units, rates compare two measurements with different units. Examples of rates would be \$200 for 8 hours of work, or 500 miles traveled per 20 gallons. Because the units are different, it is important to always include the units when discussing rates. Key words in rate problems include for, per, on, from, and in. Just as with ratios, it is important to write rates in lowest terms. A common rate in real-life situations is cost per unit, which describes how much one item/unit costs. When evaluating the cost of an item that comes in several sizes, the cost per unit rate can help buyers determine the best deal. For example, if 2 quarts of soup was sold for \$3.50 and 3 quarts was sold for \$4.60, to determine the best buy, the cost per quart should be found. $\frac{\$3.50}{2\text{ qt}} = \$1.75$ per quart, and $\frac{\$4.60}{3\text{ qt}} = \$1.53$ per quart. Therefore, the better deal would be the 3-quart option.

Rate of change problems involve calculating a quantity per some unit of measurement. Usually the unit of measurement is time. For example, meters per second is a common rate of change. To calculate this measurement, find the amount traveled in meters and divide by total time traveled. The result is the average speed over the entire time interval. Another common rate of change used in the real world is miles per hour. Consider the following problem that involves calculating an average rate of change in temperature. Last Saturday, the temperature at 1:00 a.m. was 34 degrees Fahrenheit, and at noon, the temperature had increased to 75 degrees Fahrenheit. What was the average rate of change over that time interval? The average rate of change is calculated by finding the change in temperature and dividing by the total hours elapsed. Therefore, the rate of change was equal to $\frac{75-34}{12-1} = \frac{41}{11}$ degrees per hour. This quantity rounded to two decimal places is equal to 3.73 degrees per hour.

A common rate of change that appears in algebra is the slope calculation. Given a linear equation in one variable, $y = mx + b$, the **slope**, m, is equal to:

$$\frac{rise}{run} \text{ or } \frac{change\ in\ y}{change\ in\ x}$$

In other words, slope is equivalent to the ratio of the vertical and horizontal changes between any two points on a line. The vertical change is known as the **rise**, and the horizontal change is known as the **run**. Given any two points on a line (x_1, y_1) and (x_2, y_2), slope can be calculated with the formula:

$$m = \frac{y_2 - y_1}{x_2 - x_1} = \frac{\Delta y}{\Delta x}$$

Common real-world applications of slope include determining how steep a staircase should be, calculating how steep a road is, and determining how to build a wheelchair ramp.

Many times, problems involving rates and ratios involve proportions. A proportion states that two ratios (or rates) are equal. The property of cross products can be used to determine if a proportion is true, meaning both ratios are equivalent. If $\frac{a}{b} = \frac{c}{d}$, then to clear the fractions, multiply both sides by the least common denominator, bd. This results in $ad = bc$, which is equal to the result of multiplying along both diagonals. For example, $\frac{4}{40} = \frac{1}{10}$ grants the cross product $4 \times 10 = 40 \times 1$, which is equivalent to $40 = 40$ and shows that this proportion is true.

Cross products are used when proportions are involved in real-world problems. Consider the following: If 3 pounds of fertilizer will cover 75 square feet of grass, how many pounds are needed for 375 square feet? To solve this problem, set up a proportion using two ratios. Let x equal the unknown quantity, pounds needed for 375 feet. Setting the two ratios equal to one another yields the equation:

$$\frac{3}{75} = \frac{x}{375}$$

Cross-multiplication gives $3 \times 375 = 75x$. Therefore, $1{,}125 = 75x$. Divide both sides by 75 to get $x = 15$. Therefore, 15 pounds of fertilizer are needed to cover 375 square feet of grass.

Another application of proportions involves similar triangles. If two triangles have corresponding angles with the same measurements and corresponding sides with proportional measurements, the triangles are said to be similar. If two angles are the same, the third pair of angles are equal as well because the sum of all angles in a triangle is equal to 180 degrees. Each pair of equivalent angles are known as **corresponding angles. Corresponding sides** face the corresponding angles, and it is true that corresponding sides are in proportion. For example, consider the following set of similar triangles:

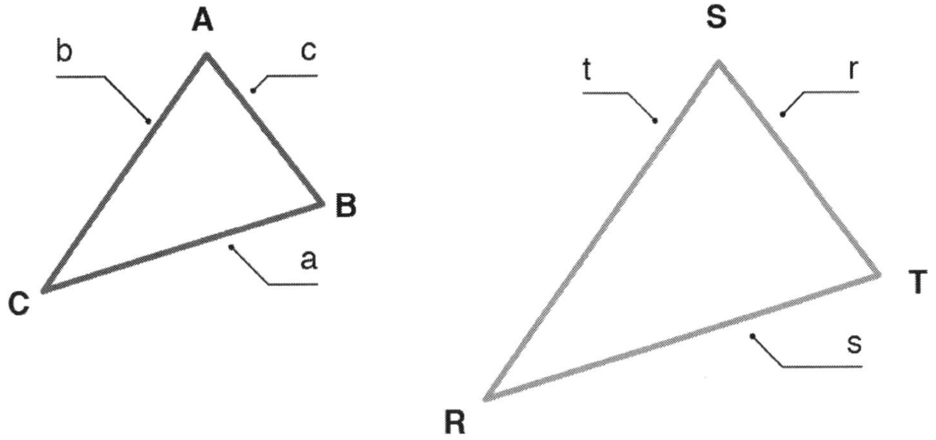

Angles A and S have the same measurement, angles C and R have the same measurement, and angles B and T have the same measurement. Therefore, the following proportion can be set up from the sides:

$$\frac{c}{r} = \frac{a}{s} = \frac{b}{t}$$

This proportion can be helpful in finding missing lengths in pairs of similar triangles. For example, if the following triangles are similar, a proportion can be used to find the missing side lengths, a and b.

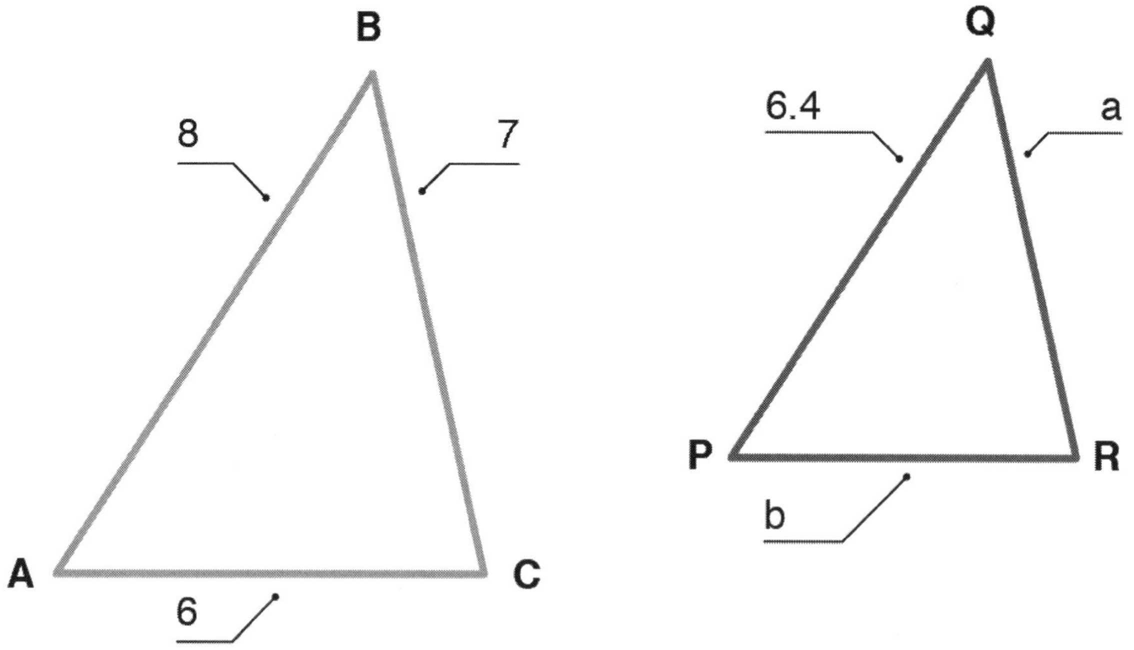

The proportions $\frac{8}{6.4} = \frac{6}{b}$ and $\frac{8}{6.4} = \frac{7}{a}$ can both be cross multiplied and solved to obtain $a = 5.6$ and $b = 4.8$.

A real-life situation that uses similar triangles involves measuring shadows to find heights of unknown objects. Consider the following problem: A building casts a shadow that is 120 feet long, and at the same time, another building that is 80 feet high casts a shadow that is 60 feet long. How tall is the first building? Each building, together with the sun rays and shadows casted on the ground, forms a triangle. They are similar because each building forms a right angle with the ground, and the sun rays form equivalent angles. Therefore, these two pairs of angles are both equal. Because all angles in a triangle add up to 180 degrees, the third angles are equal as well. Both shadows form corresponding sides of the triangle, the buildings form corresponding sides, and the sun rays form corresponding sides. Therefore, the triangles are similar, and the following proportion can be used to find the missing building length:

$$\frac{120}{x} = \frac{60}{80}$$

Cross multiply to obtain the equation $9600 = 60x$. Then, divide both sides by 60 to obtain $x = 160$. This means that the first building is 160 feet high.

Solving Real-World Problems Involving Proportions

Fractions appear in everyday situations, and in many scenarios, they appear in the real-world as ratios and in proportions. A **ratio** is formed when two different quantities are compared. For example, in a group of 50 people, if there are 33 females and 17 males, the ratio of females to males is 33 to 17. This expression can be written in the fraction form as $\frac{33}{50}$, where the denominator is the sum of females and males, or by using the ratio symbol, 33:17. The order of the number matters when forming ratios. In the same setting, the ratio of males to females is 17 to 33, which is equivalent to $\frac{17}{50}$ or 17:33. A **proportion** is an equation involving two ratios. The equation $\frac{a}{b} = \frac{c}{d}$, or $a: b = c: d$ is a proportion, for real numbers a, b, c, and d. Usually, in one ratio, one of the quantities is unknown, and cross-multiplication is used to solve for the unknown. Consider $\frac{1}{4} = \frac{x}{5}$. To solve for x, cross-multiply to obtain $5 = 4x$. Divide each side by 4 to obtain the solution $x = \frac{5}{4}$. It is also true that percentages are ratios in which the second term is 100 minus the first term. For example, 65% is 65:35 or $\frac{65}{100}$. Therefore, when working with percentages, one is also working with ratios.

Real-world problems frequently involve proportions. For example, consider the following problem: If 2 out of 50 pizzas are usually delivered late from a local Italian restaurant, how many would be late out of 235 orders? The following proportion would be solved with x as the unknown quantity of late pizzas:

$$\frac{2}{50} = \frac{x}{235}$$

Cross multiplying results in $470 = 50x$. Divide both sides by 50 to obtain $x = \frac{470}{50}$, which in lowest terms is equal to $\frac{47}{5}$. In decimal form, this improper fraction is equal to 9.4. Because it does not make sense to answer this question with decimals (portions of pizzas do not get delivered) the answer must be rounded. Traditional rounding rules would say that 9 pizzas would be expected to be delivered late. However, to be safe, rounding up to 10 pizzas out of 235 would probably make more sense.

Solving Single- and Multistep Problems Involving Percentages

Percentages are defined as parts per one hundred. To convert a decimal to a percentage, move the decimal point two units to the right and place the percent sign after the number. Percentages appear in many scenarios in the real world. It is important to make sure the statement containing the percentage is translated to a correct mathematical expression. Be aware that it is extremely common to make a mistake when working with percentages within word problems.

An example of a word problem containing a percentage is the following: 35% of people speed when driving to work. In a group of 5,600 commuters, how many would be expected to speed on the way to their place of employment? The answer to this problem is found by finding 35% of 5,600. First, change the percentage to the decimal 0.35. Then compute the product:

$$0.35 \times 5{,}600 = 1{,}960$$

Therefore, it would be expected that 1,960 of those commuters would speed on their way to work based on the data given. In this situation, the word "of" signals to use multiplication to find the answer. Another way percentages are used is in the following problem: Teachers work 8 months out of the year. What percent of the year do they work? To answer this problem, find what percent of 12 the number 8 is, because there are 12 months in a year. Therefore, divide 8 by 12, and convert that number to a percentage:

$$\frac{8}{12} = \frac{2}{3} = 0.66\overline{6}$$

The percentage rounded to the nearest tenth place tells us that teachers work 66.7% of the year. Percentage problems can also find missing quantities like in the following question: 60% of what number is 75? To find the missing quantity, turn the question into an equation. Let x be equal to the missing quantity. Therefore, $0.60x = 75$. Divide each side by 0.60 to obtain 125.

Therefore, 60% of 125 is equal to 75.

Sales tax is an important application relating to percentages because tax rates are usually given as percentages. For example, a city might have an 8% sales tax rate. Therefore, when an item is purchased with that tax rate, the real cost to the customer is 1.08 times the price in the store. For example, a $25 pair of jeans costs the customer $\$25 \times 1.08 = \$27$. If the sales tax rates is unknown, it can be determined after an item is purchased. If a customer visits a store and purchases an item for $21.44, but the price in the store was $19, they can find the tax rate by first subtracting $\$21.44 - \$19$ to obtain $2.44, the sales tax amount. The sales tax is a percentage of the in-store price. Therefore, the tax rate is $\frac{2.44}{19} = 0.128$, which has been rounded to the nearest thousandths place. In this scenario, the actual sales tax rate given as a percentage is 12.8%.

Solving Single- and Multistep Problems Involving Measurement Quantities, Units, and Unit Conversion

When working with dimensions, sometimes the given units don't match the formula, and conversions must be made. The metric system has base units of meter for length, kilogram for mass, and liter for liquid volume. This system expands to three places above the base unit and three places below. These places correspond to prefixes that each signify a specific base of 10.

The following table shows the conversions:

kilo-	**hecto-**	**deka-**	**base**	**deci-**	**centi-**	**milli-**
1,000 times the base	100 times the base	10 times the base		1/10 times the base	1/100 times the base	1/1000 times the base

To convert between units within the metric system, values with a base ten can be multiplied. The decimal can also be moved in the direction of the new unit by the same number of zeros on the number. For example, 3 meters is equivalent to 0.003 kilometers. The decimal moved three places (the same number of zeros for kilo-) to the left (the same direction from base to kilo-). Three meters is also equivalent to 3,000 millimeters. The decimal is moved three places to the right because the prefix milli- is three places to the right of the base unit.

The English Standard system used in the United States has a base unit of foot for length, pound for weight, and gallon for liquid volume. Conversions within the English Standard system are not as easy as those within the metric system because the former does not use a base ten model.

The following table shows the conversions within this system.

Length	Weight	Capacity
1 foot (ft) = 12 inches (in) 1 yard (yd) = 3 feet 1 mile (mi) = 5280 feet 1 mile = 1760 yards	1 pound (lb) = 16 ounces (oz) 1 ton = 2000 pounds	1 tablespoon (tbsp) = 3 teaspoons (tsp) 1 cup (c) = 16 tablespoons 1 cup = 8 fluid ounces (oz) 1 pint (pt) = 2 cups 1 quart (qt) = 2 pints 1 gallon (gal) = 4 quarts

When converting within the English Standard system, most calculations include a conversion to the base unit and then another to the desired unit. For example, take the following problem:

$$3 \text{ qt} = \_\_\_ \text{ c}$$

There is no straight conversion from quarts to cups, so the first conversion is from quarts to pints. There are 2 pints in 1 quart, so there are 6 pints in 3 quarts. This conversion can be solved as a proportion:

$$\frac{3 \text{ qt}}{x} = \frac{1 \text{ qt}}{2 \text{ pt}}$$

It can also be observed as a ratio 2:1, expanded to 6:3. Then the 6 pints must be converted to cups. The ratio of pints to cups is 1:2, so the expanded ratio is 6:12. For 6 pints, the measurement is 12 cups. This problem can also be set up as one set of fractions to cancel out units. It begins with the given information and cancels out matching units on top and bottom to yield the answer. Consider the following expression:

$$\frac{3 \text{ qt}}{1} \times \frac{2 \text{ pt}}{1 \text{ qt}} \times \frac{2 \text{ c}}{1 \text{ pt}}$$

It's set up so that units on the top and bottom cancel each other out:

$$\frac{3 \not{\text{qt}}}{1} \times \frac{2 \not{\text{pt}}}{1 \not{\text{qt}}} \times \frac{2 \text{ c}}{1 \not{\text{pt}}}$$

The numbers can be calculated as $3 \times 2 \times 2$ on the top and 1 on the bottom. It still yields an answer of 12 cups.

This process of setting up fractions and canceling out matching units can be used to convert between standard and metric systems. A few common equivalent conversions are $2.54 \text{ cm} = 1 \text{ in}$, $3.28 \text{ ft} = 1 \text{ m}$, and $2.205 \text{ lb} = 1 \text{ kg}$. Writing these as fractions allows them to be used in conversions. For the problem $5 \text{ meters} = \_\_\_$ ft, use the feet-to-meter conversion and start with the expression $\frac{5 \text{ m}}{1} \times \frac{3.28 \text{ ft}}{1 \text{ m}}$. The "meters" will cancel each other out, leaving "feet" as the final unit. Calculating the numbers yields 16.4 feet. This problem only required two fractions. Others may require longer expressions, but the underlying rule stays the same. When a unit in the numerator of a fraction matches a unit in the denominator, then they cancel each other out. Using this logic and the conversions given above, many units can be converted between and within the different systems.

The conversion between Fahrenheit and Celsius is found in a formula:

$$°C = (°F - 32) \times \frac{5}{9}$$

For example, to convert 78°F to Celsius, the given temperature would be entered into the formula:

$$°C = (78 - 32) \times \frac{5}{9}$$

Solving the equation, the temperature comes out to be 25.56°C. To convert in the other direction, the formula becomes:

$$°F = °C \times \frac{9}{5} + 32$$

Remember the order of operations when calculating these conversions.

Solving Unit Rate Problems

A **unit rate** is a rate with a denominator of one. It is a comparison of two values with different units where one value is equal to one. Examples of unit rates include 60 miles per hour and 200 words per minute. Problems involving unit rates may require some work to find the unit rate. For example, if Mary travels 360 miles in 5 hours, what is her speed, expressed as a unit rate? The rate can be expressed as the following fraction: $\frac{360\ miles}{5\ hours}$. The denominator can be changed to one by dividing by five. The numerator will also need to be divided by five to follow the rules of equality. This division turns the fraction into $\frac{72\ miles}{1\ hour}$, which can now be labeled as a unit rate because one unit has a value of one. Another type question involves the use of unit rates to solve problems. For example, if Trey needs to read 300 pages and his average speed is 75 pages per hour, will he be able to finish the reading in 5 hours? The unit rate is 75 pages per hour, so the total of 300 pages can be divided by 75 to find the time. After the division, the time it takes to read is four hours. The answer to the question is yes, Trey will finish the reading within 5 hours.

Given a Scatterplot, Using Linear, Quadratic, or Exponential Models to Describe How the Variables are Related

Independent and dependent are two types of variables that describe how they relate to each other. The **independent variable** is the variable controlled by the experimenter. It stands alone and is not changed by other parts of the experiment. This variable is normally represented by x and is found on the horizontal, or x-axis, of a graph. The **dependent variable** changes in response to the independent variable. It reacts to, or depends on, the independent variable. This variable is normally represented by y and is found on the vertical, or y-axis of the graph.

The relationship between two variables, x and y, can be seen on a scatterplot.

The following scatterplot shows the relationship between weight and height. The graph shows the weight as x and the height as y. The first dot on the left represents a person who is 45 kg and approximately 150 cm tall. The other dots correspond in the same way. As the dots move to the right and weight increases, height also increases. A line could be drawn through the middle of the dots to move from bottom left to top right. This line would indicate a **positive correlation** between the variables. If the variables had a **negative correlation**, then the dots would move from the top left to the bottom right.

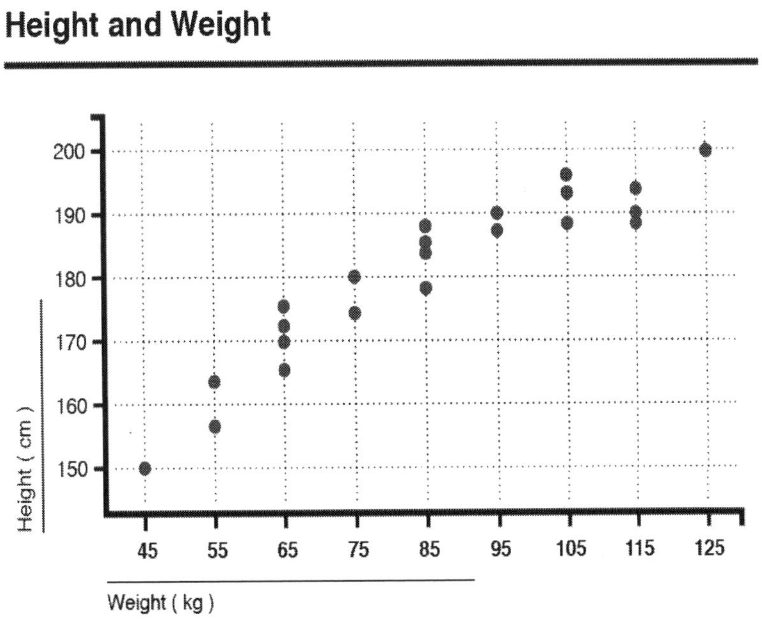

A scatterplot is useful in determining the relationship between two variables, but it is not required. Consider an example where a student scores a different grade on his math test for each week of the month. The independent variable would be the weeks of the month. The dependent variable would be the grades because they change depending on the week. If the grades trended up as the weeks passed, then the relationship between grades and time would be positive. If the grades decreased as the time passed, then the relationship would be negative. (As the number of weeks went up, the grades went down.)

The relationship between two variables can further be described as strong or weak. The relationship between age and height shows a strong positive correlation because children grow taller as they grow up. In adulthood, the relationship between age and height becomes weak, and the dots will spread out. People stop growing in adulthood, and their final heights vary depending on factors like genetics and health. The closer the dots on the graph, the stronger the relationship. As they spread apart, the relationship becomes weaker. If they are too spread out to determine a correlation up or down, then the variables are said to have no correlation.

Variables are values that change, so determining the relationship between them requires an evaluation of who changes them. If the variable changes because of a result in the experiment, then it's dependent. If the variable changes before the experiment, or is changed by the person controlling the experiment, then it's the independent variable. As they interact, one is manipulated by the other. The manipulator is the independent, and the manipulated is the dependent. Once the independent and dependent variable are determined, they can be evaluated to have a positive, negative, or no correlation.

Using the Relationship Between Two Variables to Investigate Key Features of the Graph

One way that information can be interpreted from tables, charts, and graphs is through statistics. The three most common calculations for a set of data are the mean, median, and mode. These three are called **measures of central tendency**, which are helpful in comparing two or more different sets of data. The

mean refers to the average and is found by adding up all values and dividing the total by the number of values. In other words, the mean is equal to the sum of all values divided by the number of data entries. For example, if you bowled a total of 532 points in 4 bowling games, your mean score was $\frac{532}{4} = 133$ points per game. Students can apply the concept of mean to calculate what score they need on a final exam to earn a desired grade in a class.

The **median** is found by lining up values from least to greatest and choosing the middle value. If there is an even number of values, then calculate the mean of the two middle amounts to find the median. For example, the median of the set of dollar amounts $5, $6, $9, $12, and $13 is $9. The median of the set of dollar amounts $1, $5, $6, $8, $9, $10 is $7, which is the mean of $6 and $8. The **mode** is the value that occurs the most. The mode of the data set {1, 3, 1, 5, 5, 8, 10} actually refers to two numbers: 1 and 5. In this case, the data set is bimodal because it has two modes. A data set can have no mode if no amount is repeated. Another useful statistic is range. The **range** for a set of data refers to the difference between the highest and lowest value.

In some cases, numbers in a list of data might have weights attached to them. In that case, a weighted mean can be calculated. A common application of a weighted mean is GPA. In a semester, each class is assigned a number of credit hours, its weight, and at the end of the semester each student receives a grade. To compute GPA, an A is a 4, a B is a 3, a C is a 2, a D is a 1, and an F is a 0. Consider a student that takes a 4-hour English class, a 3-hour math class, and a 4-hour history class and receives all B's. The weighted mean, GPA, is found by multiplying each grade times its weight, number of credit hours, and dividing by the total number of credit hours.

Therefore, the student's GPA is:

$$\frac{3 \times 4 + 3 \times 3 + 3 \times 4}{11} = \frac{33}{1} = 3.0.$$

The following bar chart shows how many students attend a cycle class on each day of the week. To find the mean attendance for the week, add each day's attendance together:

$$10 + 7 + 6 + 9 + 8 + 14 + 4 = 58$$

Then divide the total by the number of days, $58 \div 7 = 8.3$. The mean attendance for the week was 8.3 people. The median attendance can be found by putting the attendance numbers in order from least to greatest: 4, 6, 7, 8, 9, 10, 14, and choosing the middle number: 8 people. This set of data has no mode because no numbers repeat.

The range is 10, which is found by finding the difference between the lowest number, 4, and the highest number, 14.

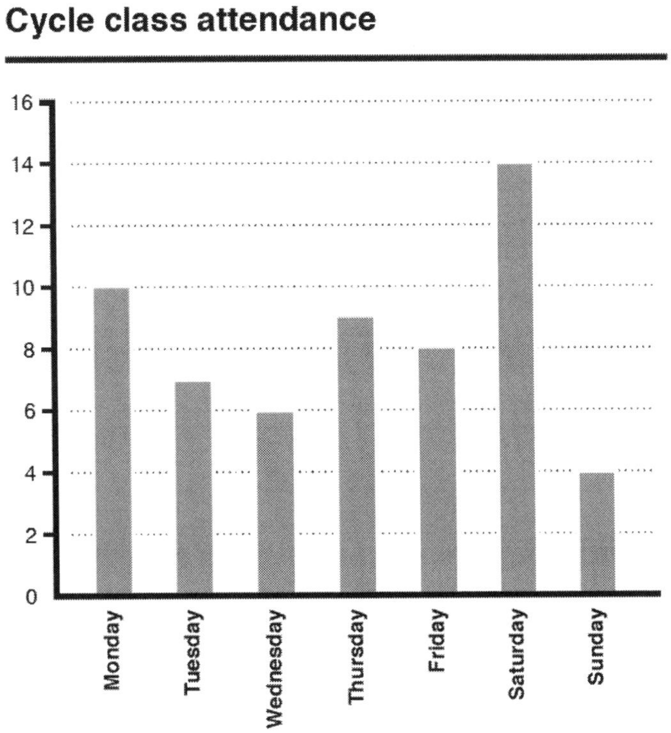

A **histogram** is a bar graph used to group data into "bins" that cover a range on the horizontal, or x-axis. Histograms consist of rectangles whose heights are equal to the frequency of a specific category. The horizontal axis represents the specific categories. Because they cover a range of data, these bins have no gaps between bars, unlike the bar graph above. In a histogram showing the heights of adult golden retrievers, the bottom axis would be groups of heights, and the y-axis would be the number of dogs in each range. Evaluating this histogram would show the height of most golden retrievers as falling within a certain range. It also provides information to find the average height and range for how tall golden retrievers may grow.

The following is a histogram that represents exam grades in a given class. The horizontal axis represents ranges of the number of points scored, and the vertical axis represents the number of students. For example, approximately 33 students scored in the 60 to 70 range.

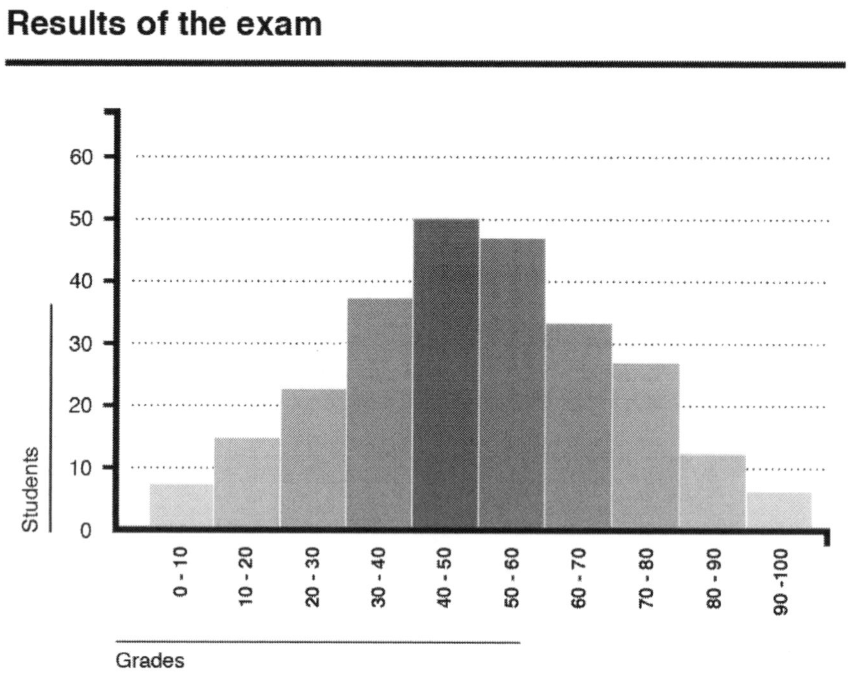

Certain measures of central tendency can be easily visualized with a histogram. If the points scored were shown with individual rectangles, the tallest rectangle would represent the mode. A **bimodal** set of data would have two peaks of equal height. Histograms can be classified as having data **skewed to the left, skewed to the right,** or **normally distributed,** which is also known as **bell-shaped.** These three classifications can be seen in the following chart:

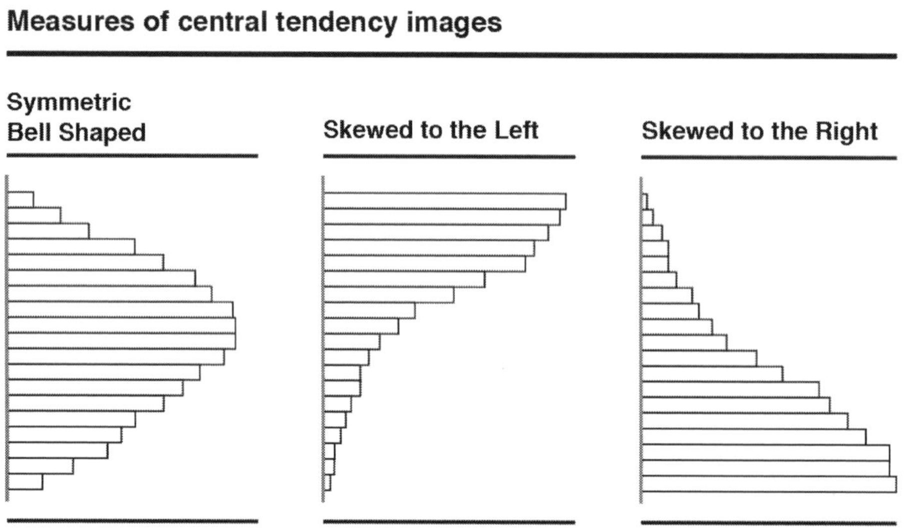

When the data is normal, the mean, median, and mode are very similar because they all represent the most typical value in the data set. In this case, the mean is typically considered the best measure of central tendency because it includes all data points. However, if the data is skewed, the mean becomes less meaningful because it is dragged in the direction of the skew. Therefore, the median becomes the best measure because it is not affected by any outliers.

The measures of central tendency and the range may also be found by evaluating information on a line graph.

In the line graph from a previous example that showed the daily high and low temperatures, the average high temperature can be found by gathering data from each day on the triangle line. The days' highs are 69, 65, 75, 74, and 81. To find the average, add them together to get 364, then divide by 5 (because there are 5 temperatures). The average high for the five days is 72.8. If 72.8 degrees is found on the graph, it will fall in the middle of all the values. The average low temperature can be found in the same way.

Given a set of data, the **correlation coefficient**, *r*, measures the association between all the data points. If two values are correlated, there is an association between them. However, correlation does not necessarily mean causation, or that one value causes the other. There is a common mistake made that assumes correlation implies causation. Average daily temperature and number of sunbathers are both correlated and have causation. If the temperature increases, that change in weather causes more people to want to catch some rays. However, wearing plus-size clothing and having heart disease are two variables that are correlated but do not have causation. The larger someone is, the more likely he or she is to have heart disease. However, being overweight does not cause someone to have the disease.

The value of the correlation coefficient is between −1 and 1, where −1 represents a perfect negative linear relationship, 0 represents no relationship between the two data sets, and 1 represents a perfect positive linear relationship. A negative linear relationship means that as *x* values increase, *y* values decrease. A positive linear relationship means that as *x* values increase, *y* values increase. The formula for computing the correlation coefficient is:

$$r = \frac{n\sum xy - (\sum x)(\sum y)}{\sqrt{n(\sum x^2) - (\sum x)^2}\sqrt{n(\sum y^2) - (y)^2}}$$

where *n* is the number of data points. The closer *r* is to 1 or −1, the stronger the correlation. A correlation can be seen when plotting data. If the graph resembles a straight line, there is a correlation.

Comparing Linear Growth with Exponential Growth

Linear growth involves a quantity, the dependent variable, increasing or decreasing at a constant rate as another quantity, the independent variable, increases as well. The graph of linear growth is a straight line. Linear growth is represented as the following equation: $y = mx + b$, where *m* is the **slope** of the line, also known as the **rate of change**, and *b* is the *y*-intercept. If the *y*-intercept is 0, then the linear growth is actually known as direct variation. If the slope is positive, the dependent variable increases as the independent variable increases, and if the slope is negative, the dependent variable decreases as the independent variable increases.

Exponential growth involves the dependent variable changing by a common ratio every unit increase. The equation of exponential growth is $y = a^x$ for $a > 0$, $a \neq 1$. The value a is known as the **base**. Consider the exponential equation $y = 2^x$. When x equals 1, y equals 2, and when x equals 2, y equals 4. For every unit increase in x, the value of the output variable doubles. Here is the graph of $y = 2^x$. Notice that as the dependent variable, y, gets very large, x increases slightly. This characteristic of this graph is why sometimes a quantity is said to be blowing up exponentially.

$y = 2^x$

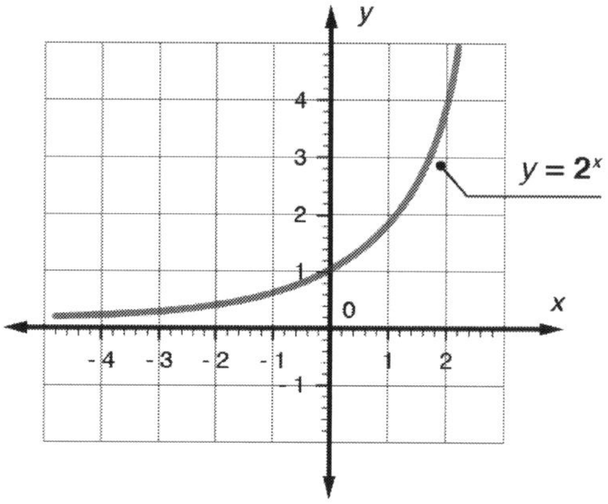

Using Two-Way Tables to Summarize Categorical Data and Relative Frequencies, and to Calculate Conditional Probability

A **two-way frequency table** displays categorical data with two variables, and it highlights relationships that exist between those two variables. Such tables are used frequently to summarize survey results and are also known as **contingency tables**. Each cell shows a count pertaining to that individual variable pairing, known as a **joint frequency**, and the totals of each row and column also are in the table. Consider the following two-way frequency table:

Distribution of the Residents of a Particular Village

	70 or older	69 or younger	Totals
Women	20	40	60
Men	5	35	40
Total	25	75	100

Table 1 shows the breakdown of ages and sexes of 100 people in a particular village. The end of each row or column displays the number of people represented by the corresponding data, and the total number of people is shown in the bottom right corner. For instance, there were 25 people age 70 or older and 60 women in the data. The 20 in the first cell shows that out of 100 total villagers, 20 were women aged 70 or older. The 5 in the cell below shows that out of 100 total villagers, 5 were men aged 70 or older.

A two-way table can also show relative frequencies by indicating the percentages of people instead of the count. If each frequency is calculated over the entire total of 100, the first cell would be 20% or 0.2. However, the relative frequencies can also be calculated over row or column totals. If row totals were used, the first cell would be:

$$\frac{20}{60} = 0.333 = 33.3\%$$

If column totals were used, the first cell would be:

$$\frac{20}{25} = 0.8 = 80\%$$

Such tables can be used to calculate **conditional probabilities**, which are probabilities that an event occurs, given another event. Consider a randomly selected villager. The probability of selecting a male 70 years old or older is $\frac{5}{100} = 0.05$ because there are 5 males over the age of 70 and 100 total villagers.

Making Inferences About Population Parameters Based on Sample Data

In statistics, a **population** contains all subjects being studied. For example, a population could be every student at a university or all males in the United States. A **sample** consists of a group of subjects from an entire population. A sample would be 100 students at a university or 100,000 males in the United States. **Inferential statistics** is the process of using a sample to generalize information concerning populations. **Hypothesis testing** is the actual process used when evaluating claims made about a population based on a sample.

A **statistic** is a measure obtained from a sample, and a **parameter** is a measure obtained from a population. For example, the mean SAT score of the 100 students at a university would be a statistic, and the mean SAT score of all university students would be a parameter.

The beginning stages of hypothesis testing starts with formulating a **hypothesis**, a statement made concerning a population parameter. The hypothesis may be true, or it may not be true. The test will answer that question. In each setting, there are two different types of hypotheses: the **null hypothesis**, written as H_0, and the **alternative hypothesis**, written as H_1. The null hypothesis represents verbally when there is not a difference between two parameters, and the alternative hypothesis represents verbally when there is a difference between two parameters. Consider the following experiment: A researcher wants to see if a new brand of allergy medication has any effect on drowsiness of the patients who take the medication. He wants to know if the average hours spent sleeping per day increases. The mean for the population under study is 8 hours, so $\mu = 8$. In other words, the population parameter is μ, the mean. The null hypothesis is $\mu = 8$ and the alternative hypothesis is $\mu > 8$. When using a smaller sample of a population, the null hypothesis represents the situation when the mean remains unaffected and the alternative hypothesis represents the situation when the mean increases. The chosen statistical test will apply the data from the sample to actually decide whether the null hypothesis should or should not be rejected.

Using Statistics to Investigate Measures of Center of Data and Analyze Shape, Center, and Spread

In statistics, measures of central tendency are measures of average. They include the mean, median, mode, and midrange of a data set. The **mean,** otherwise known as the **arithmetic average**, is found by dividing the sum of all data entries by the total number of data points. The **median** is the midpoint of the data points. If there is an odd number of data points, the median is the entry in the middle. If there is an even number of data points, the median is the mean of the two entries in the middle. The **mode** is the data point that occurs most often. Finally, the **midrange** is the mean of the lowest and highest data points. Given the spread of the data, each type of measure has pros and cons. In a **right-skewed distribution**, the bulk of the data falls to the left of the mean. In this situation, the mean is on the right of the median and the mode is on the left of the median. In a **normal distribution**, where the data are evenly distributed on both sides of the mean, the mean, median, and mode are very close to one another. In a **left-skewed distribution,** the bulk of the data falls to the right of the mean. The mean is on the left of the median and the mode is on the right of the median. Here is an example of each type of distribution:

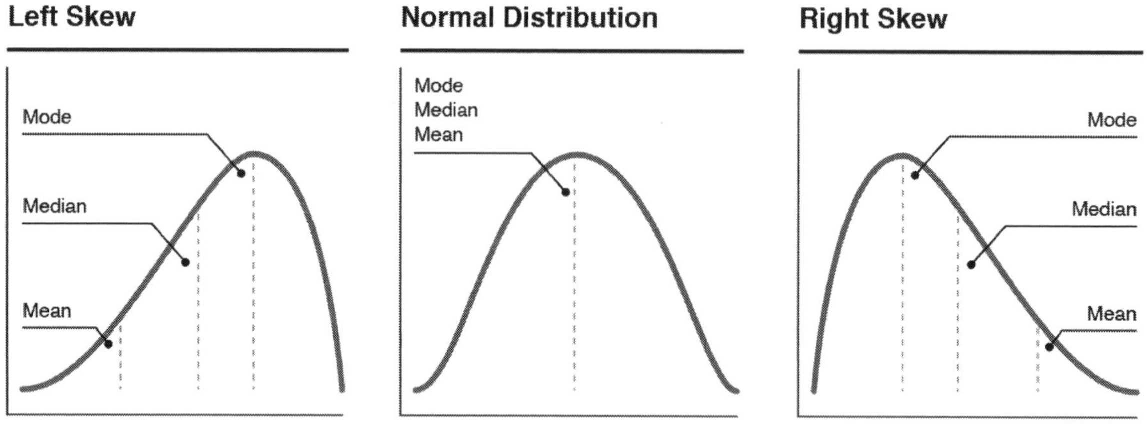

Solving Problems Involving Measures of Center and Range

A data set can be described by calculating the mean, median, and mode. These values, called measures of center, allow the data to be described with a single value that is representative of the data set.

The most common measure of center is the mean, also referred to as the average.

To calculate the mean,

- Add all data values together
- Divide by the sample size (the number of data points in the set)

The median is middle data value, so that half of the data lies below this value and half lies above it.

To calculate the median,

- Order the data from least to greatest
- The point in the middle of the set is the median
    - In the event that there is an even number of data points, add the two middle points and divide by 2

The mode is the data value that occurs most often.

To calculate the mode,

- Order the data from least to greatest
- Find the value that occurs most often

Example: Amelia is a leading scorer on the school's basketball team. The following data set represents the number of points that Amelia has scored in each game this season. Use the mean, median, and mode to describe the data.

16, 12, 26, 14, 28, 14, 12, 15, 25

Solution: Mean:

$$16 + 12 + 26 + 14 + 28 + 14 + 12 + 15 + 25 = 162$$

$$162 \div 9 = 18$$

Amelia averages 18 points per game.

Median:

12, 12, 14, 14, **15**, 16, 25, 26, 28

Amelia's median score is 15.

Mode:

12, 12, 14, 14, 15, 16, 25, 26, 28

The numbers 12 and 14 each occur twice in the data set, so this set has 2 modes: 12 and 14.

The **range** is the difference between the largest and smallest values in the set. In the example above, the range is:

$$28 - 12 = 16$$

Evaluating Reports to Make Inferences, Justify Conclusions, and Determine Appropriateness of Data Collection Methods

For researchers to make valid conclusions about population characteristics and parameters, the sample used to compare must be random. In a **random sample**, every member of the population must have an equal chance of being selected. In this setting, the sample is **unbiased** and is said to be a good representation of the population. If a sample is selected in an inappropriate manner, it is said to be **biased.** A sample can be biased if, for example, some subjects were more likely to be chosen than others.

In order to have unbiased samples, the four main sampling methods used tend to be random, systematic, stratified, and cluster sampling.

Random sampling occurs when, given a sample size *n*, all possible samples of that size are equally likely to be chosen. Random numbers from calculators are typically used in this setting. Each member of a population is paired with a number, and then a set of random numbers is generated. Each person paired with one of those random numbers is selected. A **systematic sample** is when every fourth, seventh, tenth, etc., person from a population is selected to be in a sample. A **stratified sample** is when the population is divided into subgroups, or strata, using a characteristic, and then members from each stratum are randomly selected. For instance, university students could be divided into age groups and then selected from each age group. Finally, a **cluster sample** is when a sample is used from an already selected group, like city block or zip code. These four methods are used most frequency because they are most likely to yield unbiased results.

Once an unbiased sample is found, data need to be collected. Common data collection methods include surveys with unbiased questions, words that are not confusing, questions that do not use double negatives, and questions that do not contain compound sentences that ask two questions at once. The simpler verbiage the better when formulating these questions.

Passport to Advanced Math

Creating a Quadratic or Exponential Function or Equation

When given data in ordered pairs, choosing an appropriate function or equation to model the data is important. Besides linear relationships, other common relationships that exist are quadratic and exponential. A helpful way to determine what type of function to use is to find the difference between consecutive dependent variables. Basically, find pairs of ordered pairs where the *x*-values increase by 1, and take the difference of the *y*-values. If the differences in the *y*-values are always the same value, then the function is **linear**. If the differences in the *y*-values when the *x*-values increase by 1 are not the same, the function could be quadratic or exponential. If the differences are not the same, find differences of those differences. If consecutive differences are the same, then the function is **quadratic**. If consecutive differences are not the same, try taking ratios of consecutive *y*-values. If the ratios are the same, the data have an exponential relationship and an exponential function should be used.

For example, the ordered pairs (1, 4), (2, 6), (3 ,8), and (4,10) have a linear relationship because the difference in *y*-values is 2 for every increase in *x* of 1. The ordered pairs (1, 0), (2, 3), (3, 10), and (4, 21) have a nonlinear relationship. The first differences in *y*-values are 3, 7, and 11, however, consecutive second differences are both 4. Third, the ordered pairs (1, 10), (2, 30), (3, 90), and (4, 270) have an exponential relationship. Taking ratios of consecutive *y*-values leads to a common ratio of 4.

The general form of a **quadratic equation** is $y = ax^2 + bx + c$, and its vertex form is $y = a(x - h)^2 + k$, with vertex (h, k). If the vertex and one other point are known, the vertex form should be used to solve for *a*. If three points, not the vertex, are known, the general form should be used. The three points create a system of three equations in three unknowns that can be solved for.

The general form of an **exponential function** i:

$$y = b \times a^x$$

where *a* is the base and *b* is the y-intercept.

Determining the Most Suitable Form of an Expression or Equation

When given a problem, it is necessary to determine the best form of an expression or equation to use, given the context. Usually this involves some algebraic manipulation. If an equation is given, the simplest form of the equation is best. Simplifying involves using the distributive property, collecting like terms, etc., If an equation is needed to be solved, properties involving performing the same operation on both sides of the equation must be used. For instance, if a number is added to one side of the equals sign, it must be added to the other side as well. This maintains a true equation.

If an expression is given, simplifying can only involve properties allowing to rewrite the expression as an equivalent form. If there is no equals sign, mathematical operations cannot be performed on the expression, unless it is a rational expression. A **rational expression** can be written in the form of a fraction, in which the numerator and denominator are both polynomials and the denominator is not equal to zero. Rational expressions can always be multiplied times a form of 1. For example, consider the following rational expression involving radicals: $\frac{2}{\sqrt{2}}$. It is incorrect to write a fraction with a root in the denominator, and therefore the expression must be rationalized. Multiply the fraction times $\frac{\sqrt{2}}{\sqrt{2}}$, a form of 1. This results in:

$$\frac{2}{\sqrt{2}} \times \frac{\sqrt{2}}{\sqrt{2}} = \frac{2\sqrt{2}}{\sqrt{4}} = \frac{2\sqrt{2}}{2} = \sqrt{2}$$

which is the most suitable form of the expression.

Creating Equivalent Expressions Involving Rational Exponents and Radicals

Writing radical expressions into equivalent forms involving rational exponents can help in simplifying complex radical expressions. The rule that helps this conversion is:

$$\sqrt[n]{x^m} = x^{\frac{m}{n}}. \text{ If } m = 1$$

the rule is simply $\sqrt[n]{x} = x^{\frac{1}{n}}$. For instance, consider the following expression: $\sqrt[4]{x}\sqrt[2]{y}$.

It can be written as one radical expression, but first it needs to be converted to an equivalent expression using rational expressions. The equivalent expression is $x^{\frac{1}{4}}y^{\frac{1}{2}}$. The goal is to have one radical, which means one index *n*, so a common denominator of the exponents must be found. The common denominator is 4, so an equivalent expression is $x^{\frac{1}{4}}y^{\frac{2}{4}}$. The exponential rule:

$$a^m b^m = (ab)^m$$

can be used to, in a sense, factor out a $\frac{1}{4}$ out of both exponents. This process results in the expression $(xy^2)^{\frac{1}{4}}$, and its equivalent radical form is $\sqrt[4]{xy^2}$. Converting to rational exponents has allowed the entire expression to be written as one radical.

Another type of problem could involve going in the opposite direction: starting with rational exponents and using an equivalent radical form to simplify the expression. For instance, $32^{\frac{1}{5}}$ might not seem obviously equal to 2. However, putting it in is equivalent radical form $\sqrt[5]{32}$ shows that it is equivalent to the fifth root of 32, which is 2.

Creating an Equivalent Form of an Algebraic Expression

Two algebraic expressions are equivalent if, even though they look different, they represent the same expression. Therefore, plugging in the same values into the variables in each expression will result in the same result in both expressions. To obtain an equivalent form of an algebraic expression, laws of algebra must be followed. For instance, addition and multiplication are both commutative and associative. Therefore, terms in an algebraic expression can be added in any order and multiplied in any order. For instance,

$$4x + 2y \text{ is equivalent to}$$

$$2y + 4x \; and \; y \times 2 + x \times 4.$$

Also, the distributive law allows a number to be distributed throughout parentheses, as in the following:

$$a(b + c) = ab + ac$$

The two expressions on both sides of the equals sign are equivalent. Also, collecting like terms is important when working with equivalent forms. The simplest version of an expression is always the one easiest to work with, so all like terms (those with the same variables raised to the same powers) must be combined.

Note that an expression is not an equation, and therefore expressions cannot be multiplied times numbers, divided by numbers, or have numbers added to them or subtracted from them and still have equivalent expressions. These processes can only happen in equations when the same step is performed on both sides of the equals sign.

Solving a Quadratic

Given a quadratic equation in standard form:

$$ax^2 + bx + c = 0$$

with constants *a*, *b*, and *c*, such that $c \neq 0$, it can have either two real solutions, one real solution, or two complex solutions of the form $a + bi$ (no real solutions). The number of solutions is determined using its determinant $b^2 - 4ac$. If the determinant is positive, there are two real solutions. If the determinant is negative, there are no real solutions. If the determinant is equal to 0, there is one real solution. For example, given the quadratic equation:

$$2x^2 - x + 4 = 0$$

Its determinant is:

$$(-1)^2 - 4(2)(4) = 1 - 32 = -31$$

which is less than 0. Therefore, it has two complex solutions.

There are a number of ways to solve a quadratic equation. The first way is through **factoring.** If the equation is in standard form and the polynomial can be factored, set each factor equal to 0 and solve. This can be done because if $ab = 0$, either $a = 0, b = 0$, or both are equal to 0. For example:

$$x^2 - 7x + 10 = (x - 5)(x - 2)$$

Therefore, the solutions of:

$$x^2 - 7x + 10 = 0$$

are those that satisfy both $x - 5 = 0$ and $x - 2 = 0$, or $x = 5$ and 2. This is the simplest method to solve quadratic equations; however, not all quadratic polynomials can be factored, so this method does not work for all quadratic equations.

Another method is through **completing the square**. The polynomial $x^2 + 10x - 9$ cannot be factored, so complete the square in the equation $x^2 + 10x - 9 = 0$ to find its solutions. First, add 9 to both sides, resulting in:

$$x^2 + 10x = 9$$

Then, divide the *x*-coefficient by 2, square it, and add it to both sides of the equation. In this example, $\left(\frac{10}{2}\right)^2 = 25$ is added to both sides of the equation to obtain:

$$x^2 + 10x + 25 = 9 + 25 = 34$$

The polynomial, which is now a perfect square trinomial, can then be factored into:

$$(x + 5)^2 = 34$$

Finally, solving for *x* involves first taking the square root of both sides and then subtracting 5 from both sides. This process leads to the two solutions:

$$x = \pm\sqrt{34} - 5$$

This method always works for any quadratic equation.

The final method of solving a quadratic equation is to use the **quadratic formula.** Given a quadratic equation in standard form, $ax^2 + bx + c = 0$, its solutions always can be found using the formula:

$$x = \frac{-b \pm \sqrt{b^2 - 4ac}}{2a}$$

This method, like completing the square, can always be used.

Adding, Subtracting, and Multiplying Polynomial Equations

When working with polynomials, **like terms** are terms that contain exactly the same variables with the same powers. For example, x^4y^5 and $9x^4y^5$ are like terms. The coefficients are different, but the same variables are raised to the same powers. When adding polynomials, only terms that are like can be added.

When adding two like terms, just add the coefficients and leave the variables alone. This process uses the distributive property. For example:

$$x^4y^5 + 9x^4y^5 = (1 + 9)x^4y^5 = 10x^4y^5$$

Therefore, when adding two polynomials, simply add the like terms together. Unlike terms cannot be combined.

Subtracting polynomials involves adding the opposite of the polynomial being subtracted. Basically, the sign of each term in the polynomial being subtracted is changed, and then the like terms are combined because it is now an addition problem. For example, consider the following:

$$6x^2 - 4x + 2 - (4x^2 - 8x + 1).$$

Add the opposite of the second polynomial to obtain:

$$6x^2 - 4x + 2 + (-4x^2 + 8x - 1)$$

Then, collect like terms to obtain:

$$2x^2 + 4x + 1$$

Multiplying polynomials involves using the product rule for exponents that:

$$b^m b^n = b^{m+n}$$

Basically, when multiplying expressions with the same base, just add the exponents. Multiplying a monomial times a monomial involves multiplying the coefficients together and then multiplying the variables together using the product rule for exponents. For instance:

$$8x^2y \times 4x^4y^2 = 32x^6y^3$$

When multiplying a monomial times a polynomial that is not a monomial, use the distributive property to multiply each term of the polynomial times the monomial. For example:

$$3x(x^2 + 3x - 4) = 3x^3 + 9x^2 - 12x$$

Finally, multiplying two polynomials when neither one is a monomial involves multiplying each term of the first polynomial times each term of the second polynomial. There are some shortcuts, given certain scenarios. For instance, a binomial times a binomial can be found by using the **FOIL (Firsts, Outers, Inners, Lasts)** method shown here:

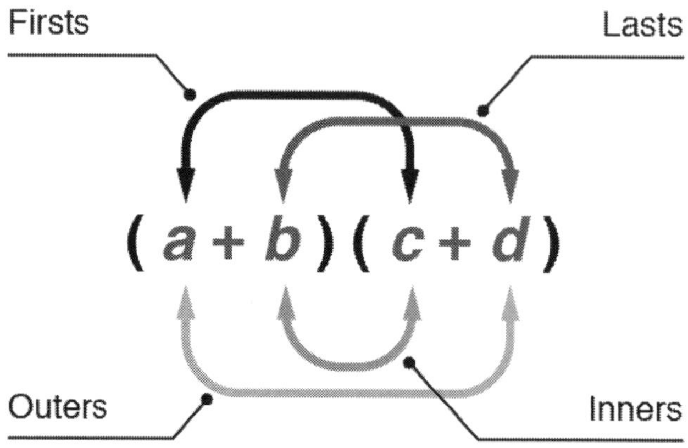

Finding the product of a sum and difference of the same two terms is simple because if it was to be foiled out, the outer and inner terms would cancel out. For instance:

$$(x + y)(x - y) = x^2 + xy - xy - y^2$$

Finally, the square of a binomial can be found using the following formula:

$$(a \pm b)^2 = a^2 \pm 2ab + b^2$$

Solving an Equation in One Variable that Contains Radicals or Contains the Variable in the Denominator of a Fraction

A **radical equation** is an equation that contains a variable in the **radicand,** which is the expression under the root. The radical can be a square root, a cube root, or a higher root. To solve an equation containing a root, arrange the terms so that one radical is by itself on one side of the equals sign. Then raise both sides of the equation to the value of the root. For example, square both sides if the root contains a square root, and cube both sides if the root contains a cube root. Then, solve the resulting equation. If the equation still contains a radical, those steps must be completed again to remove the equation of all radicals. Finally, it is crucial that all solutions are checked in the original equation. Some solutions to this equation, once the radicals are removed, might not be solutions to the original radical equation. Consider the following radical equation:

$$\sqrt{3x + 1} - \sqrt{x + 4} = 1$$

Add the second radical to both sides, and then square both sides to obtain:

$$3x + 1 = x + 4 + 2\sqrt{x + 4} + 1.$$

Next, collect like terms and isolate the radical to obtain:

$$4x^2 - 16x + 16 = 4(x + 4).$$

This simplifies into the quadratic equation $4x^2 - 20x$, which can be solved using factoring:

$$4x(x - 5) = 0.$$

So, it has solutions, $x = 0$ and $x = 5$. Both values must be checked into the original radical equation. because $x = 0$ does not check, it is not a real solution and is called an **extraneous solution**. However, $x = 5$ is a solution.

If an equation contains a variable in the denominator of a fraction, it is known as a **rational equation**. Anything, that when plugged into the equation, contains a zero denominator, cannot be a solution of the equation. To solve a rational equation, multiply both sides of the equation by the LCD (least common denominator) of all of the terms in the equation. Then, solve the resulting equation, making sure that the solutions do not cause any term to have a zero denominator in the original equation.

Here is an example of solving the rational equation:

$$\frac{5}{x} - \frac{1}{3} = \frac{1}{x}$$

$$3x \times \left(\frac{5}{x} - \frac{1}{3}\right) = 3x \times \left(\frac{1}{x}\right)$$ Multiply both sides by the LCD

$$3x \times \frac{5}{x} - 3x \times \frac{1}{3} = 3x \times \frac{1}{x}$$ Distribute

$$15 - x = 3$$

$$-x = -12$$ Simplify, and then solve

$$x = 12$$

Both sides of the equation were initially multiplied times the LCD, $3x$. Note that the only number that could cause a problem as a solution would be $x = 0$, because it would create a 0 in the denominator.

Solving a System of One Linear Equation and One Quadratic Equation

The graph of a linear equation is a line, and the graph of a quadratic equation is a parabola. Together, they form a system of a linear and quadratic equation. Here is an example of such a system:

$$\begin{cases} y = -2x + 3 \\ y = x^2 - 6x + 3 \end{cases}$$

Its solution consists of the points in which the two graphs intersect. Therefore, systems of equations involving a linear and quadratic equation can be solved graphically by physically locating those points, as in the following graphic:

The Graph of the System has Two Solutions:

$$\begin{cases} y = -2x + 3 \\ y = x^2 - 6x + 3 \end{cases}$$

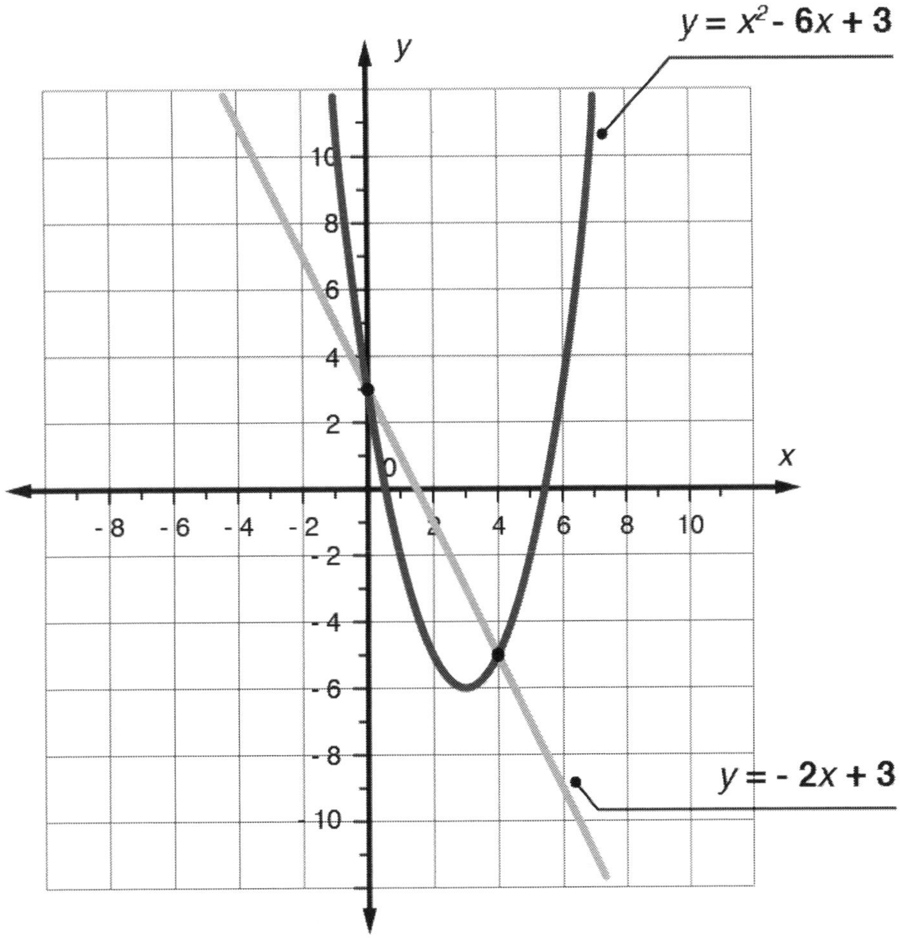

In this example, there are two points at which the graphs intersect, so there are two solutions. However, there could also be only one solution, as seen in the following example:

A System of Equations with a Single Solution

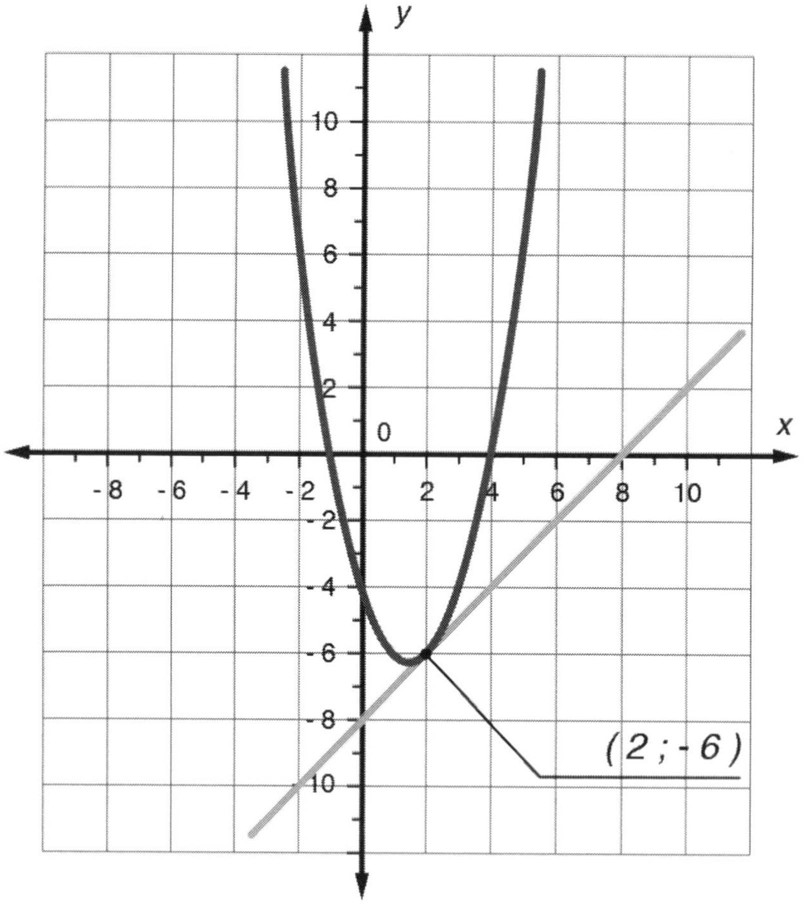

Third, it could be true that the graphs do not intersect, and in this case, there is no solution.

The solutions can also be found algebraically. In this case, solve for *y* in both equations and then set them equal to one another, as an equation in one variable, *x*. Then, put it in **standard form**, meaning move all terms to one side of the equals sign so that 0 is on the other side. This results in a quadratic equation that needs to be solved. It can be solved using factoring, completing the square, or by applying the quadratic formula. This results in either one solution (one point of intersection), two solutions (two points of intersection), or complex number solutions (no solution, no points of intersection). Then, substitute the *x*-values back into the linear equation to find the corresponding *y*-values. This results in the entire ordered pair solutions of the system.

The system above already has both equations equal to *y*, so they can be set equal to one another as:

$$x^2 - 6x + 3 = -2x + 3$$

Moving all terms to the left side results in:

$$x^2 - 4x = 0$$

Factoring the expression on the left gives:

$$x(x - 4) = 0$$

Setting both factors equal to 0 results in $x = 0$ and $x = 4$.

Plugging both *x*-values into the linear equation results in $y = 3$ and $y = -5$, respectively. Therefore, the ordered pair solutions are (0, 3) and (4, -5). Both of these ordered pairs satisfy both equations in the original system.

Rewriting Simple Rational Expressions

A **rational expression** is a fraction or a ratio in which both the numerator and denominator are polynomials that are not equal to zero. A **polynomial** is a mathematical expression containing the sum and difference of one or more terms that are constants multiplied times variables raised to positive powers. Here are some examples of rational expressions:

$$\frac{2x^2 + 6x}{x}$$

$$\frac{x-2}{x^2-6x+8}$$

$$\frac{x + 2}{x^3 - 1}$$

Such expressions can be simplified using different forms of division. The first example can be simplified in two ways. First, because the denominator is a monomial, the expression can be split up into two expressions:

$$\frac{2x^2}{x} + \frac{6x}{x}$$

and then simplified using properties of exponents as $2x + 6$. It also can be simplified using factoring and then crossing common factors out of the numerator and denominator. For instance, it can be written as

$$\frac{2x(x + 3)}{x} = 2(x + 3) = 2x + 6$$

The second expression above can also be simplified using factoring. It can be written as:

$$\frac{x - 2}{(x - 2)(x - 4)} = \frac{1}{x - 4}$$

Finally, the third example can only be simplified using long division, as there are no common factors in the numerator and denominator. First, divide the first term of the denominator by the first term of the numerator, then write that in the quotient. Then, multiply the divisor times that number and write it below the dividend. Subtract, bring down the next term from the dividend, and continue that process with the next first term and first term of the divisor. Continue the process until every term in the divisor is accounted for.

Here is the actual long division:

Simplifying Expressions Using Long Division

	x^2	$-2x$	$+4$	
$x+2$	x^3			-1
	x^3	$+2x^2$		
		$-2x^2$		-1
		$-2x^2$	$-4x$	
			$4x$	-1
			$4x$	$+8$
				-9

Interpreting Parts of Nonlinear Expressions in Terms of Their Context

If a quantity increases or decreases at a constant rate as another quantity increases, then this idea is represented as a **linear expression**, and the graph of such a relationship is a straight line. All other relationships are nonlinear, and nonlinear expressions must be used as mathematical representations of such instances.

Common nonlinear relationships that exist between two quantities are **inverse variation equations**, which are represented by equations such as $y = \frac{k}{x}$ and $y = \frac{k}{x^2}$ with constants k, quadratic equations of the form:

$$y = ax^2 + bx + c$$

and exponential equations of the form $y = a^x$.

Inverse variation situations arise when, as one quantity increases, the other quantity decreases in proportion. For instance, as a person increases the speed of a car she is driving, the time it takes to reach the destination decreases. This is a nonlinear relationship regarding inverse variation.

Recall that quadratic equations are used to model something shaped like a parabola. For instance, if a ball is thrown into the air, it travels higher and higher and eventually slows down to reach its highest point, then stops dropping at a faster and faster rate. A quadratic equation must be used to tell the position of the ball given the amount of time since the ball was thrown. This relationship is nonlinear.

Finally, an **exponential equation** is used to model something with exponential growth or decay. If something grows exponentially, such as compound interest, the amount is multiplied times a growth factor for every increase in *x*. If something decays exponentially, the amount is multiplied times a factor between 0 and 1 for every increase in *x*. When a population is declining, an exponential decay equation can be used to represent the situation.

Understanding the Relationship Between Zeros and Factors of Polynomials

A **polynomial** is a mathematical expression containing the sum and difference of one or more terms that are constants multiplied times variables raised to positive powers. A **polynomial equation** is a polynomial set equal to another polynomial, or in standard form, a polynomial is set equal to zero. A **polynomial function** is a polynomial set equal to y. For instance, $x^2 + 2x - 8$ is a polynomial, $x^2 + 2x - 8 = 0$ is a polynomial equation, and $y = x^2 + 2x - 8$ is the corresponding polynomial function. To solve a polynomial equation, the x-values in which the graph of the corresponding polynomial function crosses the x-axis are sought.

These coordinates are known as the **zeros** of the polynomial function because they are the coordinates in which the *y*-coordinates are 0. One way to find the zeros of a polynomial is to find its factors, then set each individual factor equal to 0, and solve each equation to find the zeros. A **factor** is a linear expression, and to completely factor a polynomial, the polynomial must be rewritten as a product of individual linear factors. The polynomial listed above can be factored as $(x + 4)(x - 2)$. Setting each factor equal to zero results in the zeros:

$$x = -4 \ and \ x = 2$$

Here is the graph of the zeros of the polynomial:

The Graph of the Zeros of $x^2 + 2x - 8 = 0$

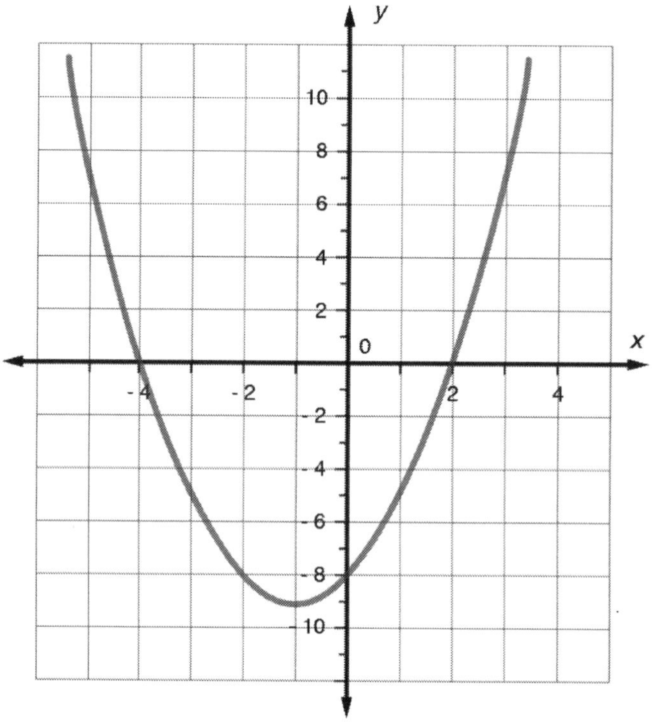

Understanding a Nonlinear Relationship Between Two Variables

Recall that a linear relationship exists between two variables if they are proportional to each other. Basically, if one quantity increases, the other quantity increases or decreases at a constant rate. The graph of a linear relationship is a straight line. If the line goes through the **origin,** the point (0, 0), then there is direct variation between the two quantities and the equation for direct variation is $y = kx$, where k is known as the constant of variation.

Recall also that a nonlinear relationship exists between two variables if an increase in one quantity does not correspond with a constant change in the other quantity. The graphs of nonlinear relationships are not straight lines. Often, the graphs of nonlinear relationships involve curves, where the change in the variable is predictable, but not constant.

A common nonlinear relationship between two variables involves inverse variation where one quantity varies inversely with respect to another. The equation for inverse variation is $y = \frac{k}{x}$, where k is still known as the constant of variation. Here is the graph of the curve $y = \frac{3}{x}$:

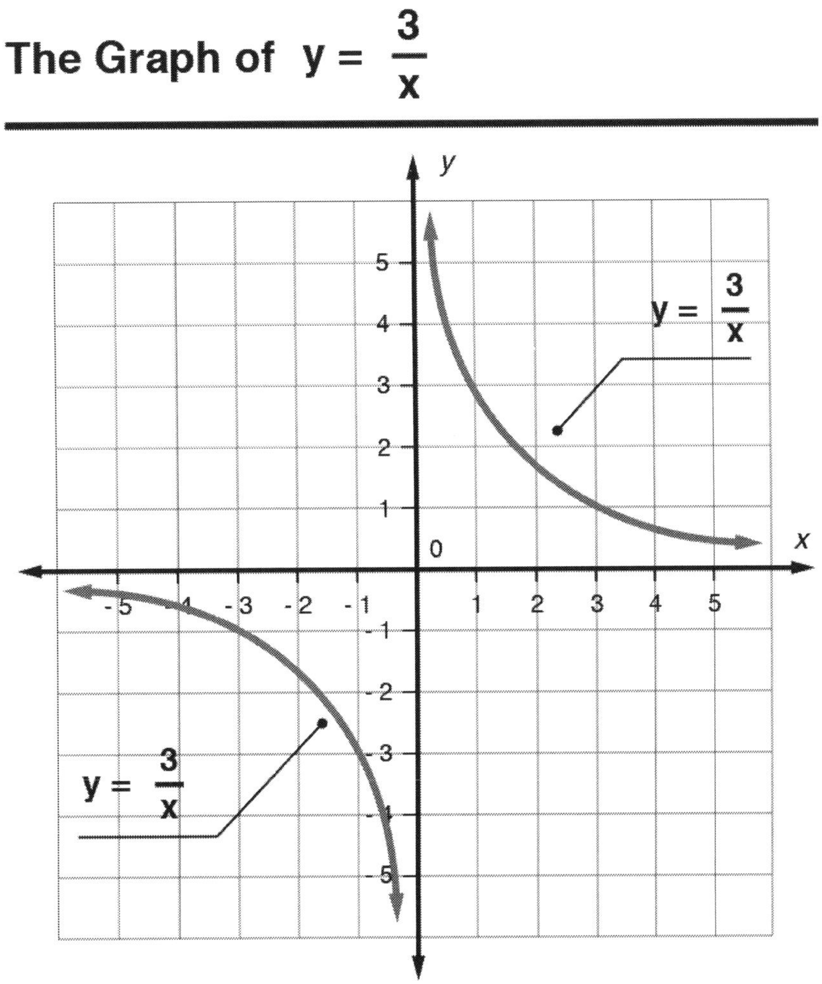

Other common nonlinear functions involve polynomial functions. The squaring function $f(x) = x^2$ can be seen here:

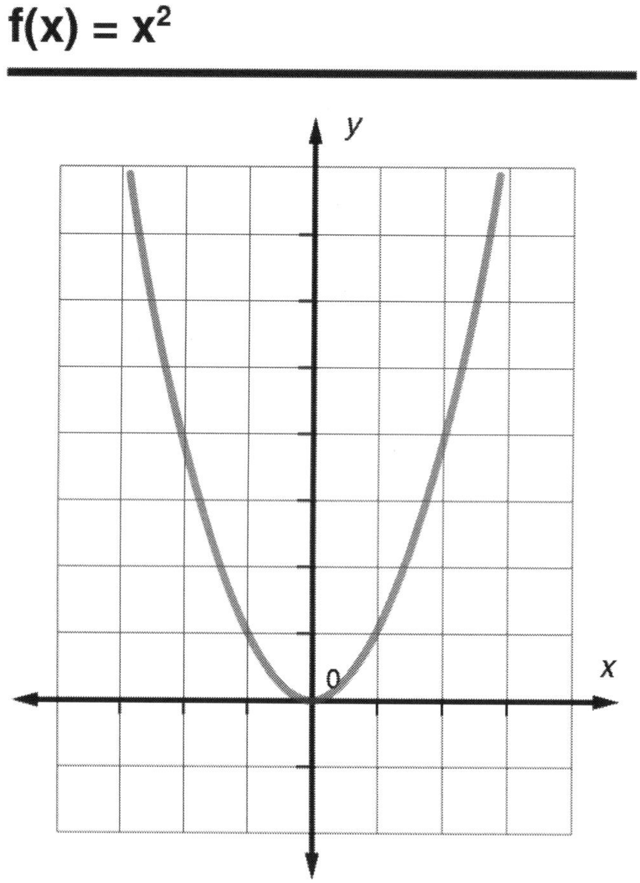

Notice that as the independent variable x increases when $x > 0$, the dependent variable y also increases. However, y does not increase at a constant rate.

Using Function Notation, and Interpreting Statements Using Function Notation

A **relation** is any set of ordered pairs (x, y). The values listed first in the ordered pairs, known as the x-coordinates, make up the domain of the relation. The values listed second, known as the y-coordinates, make up the range. A relation in which every member of the domain corresponds to only one member of the range is known as a **function.** A function cannot have a member of the domain corresponding to two members of the range. Functions are most often given in terms of equations instead of ordered pairs. For instance, here is an equation of a line: $y = 2x + 4$. In function notation, this can be written as:

$$f(x) = 2x + 4$$

The expression $f(x)$ is read "f of x" and it shows that the inputs, the x-values, get plugged into the function and the output is $y = f(x)$. The set of all inputs are in the domain, and the set of all outputs are in the range.

The x-values are known as the **independent variables** of the function and the y-values are known as the **dependent variables** of the function. The y-values depend on the x-values. For instance, if $x = 2$ is plugged into the function shown above, the y-value depends on that input.

$$f(2) = 2 \times 2 + 4 = 8$$

Therefore, $f(2) = 8$, which is the same as writing the ordered pair (2, 8). To graph a function, graph it in equation form. Therefore, replace $f(x)$ with y and plot ordered pairs.

Due to the definition of a function, the graph of a function cannot have two of the same x-components paired to different y-component. For example, the ordered pairs (3, 4) and (3, -1) cannot be in a valid function. Therefore, all graphs of functions pass the **vertical line test**. If any vertical line intersects a graph in more than one place, the graph is not that of a function. For instance, the graph of a circle is not a function because one can draw a vertical line through a circle and intersect the circle twice. Common functions include lines and polynomials, which pass the vertical line test.

Using Structure to Isolate or Identify a Quantity of Interest

When solving equations, it is important to note which quantity must be solved for. This quantity can be referred to as the **quantity of interest**. The goal of solving is to isolate the variable in the equation using logical mathematical steps. The **addition property of equality** states that the same real number can be added to both sides of an equation and equality is maintained. Also, the same real number can be subtracted from both sides of an equation to maintain equality. Second, the **multiplication property of equality** states that the same nonzero real number can multiply both sides of an equation, and still, equality is maintained. Because division is the same as multiplying times a reciprocal, an equation can be divided by the same number on both sides as well.

When solving inequalities, the same ideas are used. However, when multiplying by a negative number on both sides of an inequality, the inequality symbol must be flipped in order to maintain the logic. The same is true when dividing both sides of an inequality by a negative number.

Basically, in order to isolate a quantity of interest in either an equation or inequality, the same thing must be done to both sides of the equals sign, or inequality symbol, to keep everything mathematically correct.

Additional Topics in Math

Solving Problems Using Volume Formulas

Perimeter and **area** are geometric quantities that describe objects' measurements. **Perimeter** is the distance around an object. The perimeter of an object can be found by adding the lengths of all sides. Perimeter may be used in problems dealing with lengths around objects such as fences or borders. It may also be used in finding missing lengths or working backwards. If the perimeter is given, but a length is missing, use subtraction to find the missing length. Given a square with side length s, the formula for perimeter is $P = 4s$. Given a rectangle with length l and width w, the formula for perimeter is $P = 2l + 2w$. The perimeter of a triangle is found by adding the three side lengths, and the perimeter of a trapezoid is found by adding the four side lengths. The units for perimeter are always the original units of length, such as meters, inches, miles, etc. When discussing a circle, the distance around the object is referred to as its **circumference,** not perimeter. The formula for the circumference of a circle is $C = 2\pi r$, where r represents the radius of the circle. This formula can also be written as $C = d\pi$, where d represents the diameter of the circle.

Area is the two-dimensional space covered by an object. These problems may include the area of a rectangle, a yard, or a wall to be painted. Finding the area may require a simple formula or multiple formulas used together. The units for area are square units, such as square meters, square inches, and square miles. Given a square with side length s, the formula for its area is $A = s^2$. Some other formulas for common shapes are shown below:

Shape	Formula	Graphic
Rectangle	$Area = length \times width$	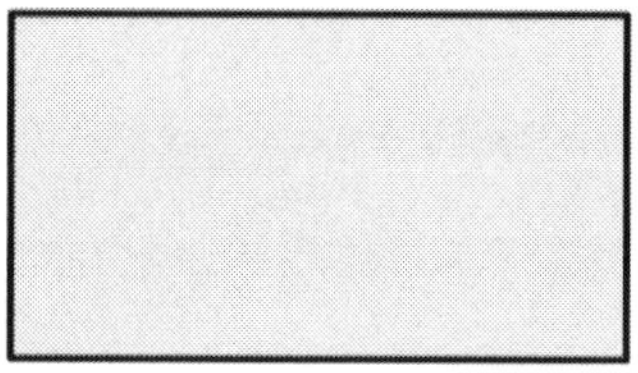
Triangle	$Area = \frac{1}{2} \times base \times height$	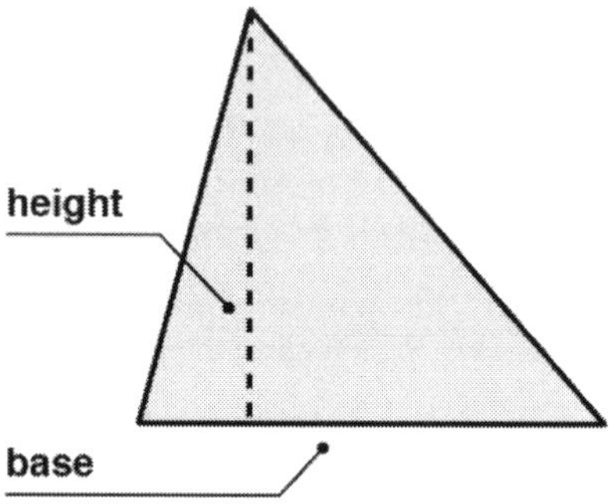
Circle	$Area = \pi \times radius^2$	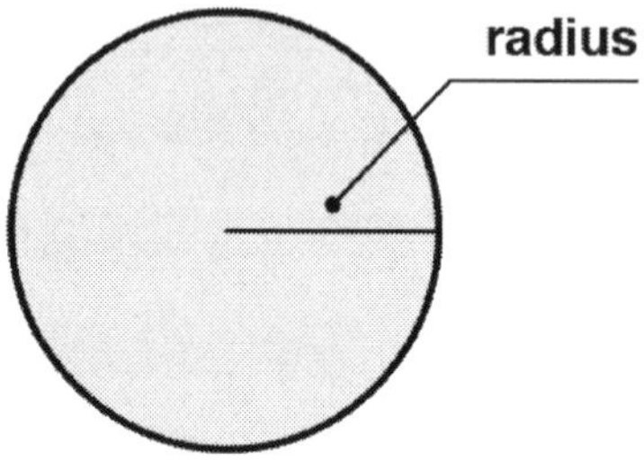

The following formula, not as widely used as those shown above, but very important, is the area of a trapezoid:

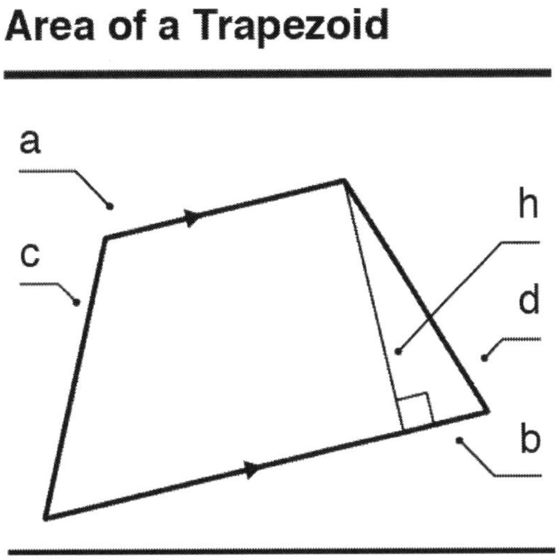

$$A = \frac{1}{2}(a + b)h$$

To find the area of the shapes above, use the given dimensions of the shape in the formula. Complex shapes might require more than one formula. To find the area of the figure below, break the figure into two shapes. The rectangle's dimensions are 6 cm by 7 cm. The triangle has a base of 4 cm and a height of 6 cm. Plug the dimensions into the rectangle formula: $A = 6 \times 7$. Multiplication yields an area of 42 cm^2. The triangle's area can be found using the formula:

$$A = \frac{1}{2} \times 4 \times 6$$

Multiplication yields an area of 12 cm^2. Add the two areas to find the total area area of the figure, which is 54 cm^2.

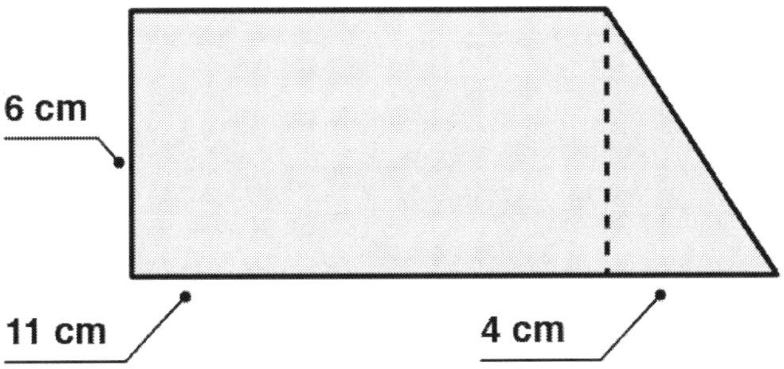

Instead of combining areas, some problems may require subtracting them, or finding the difference.

To find the area of the shaded region in the figure below, determine the area of the whole figure. Then subtract the area of the circle from the whole.

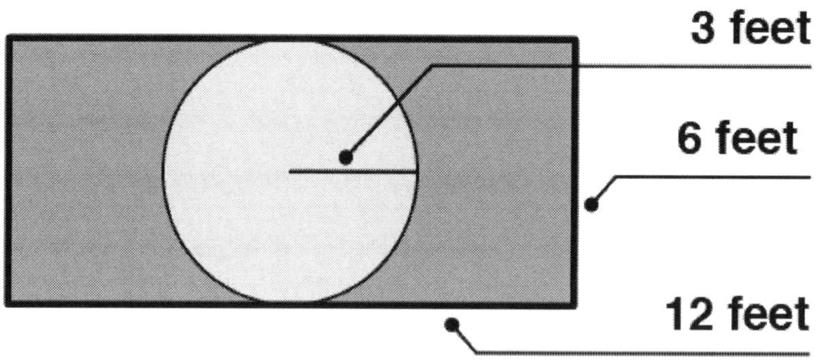

The following formula shows the area of the outside rectangle:

$$A = 12 \times 6 = 72 \text{ ft}^2$$

The area of the inside circle can be found by the following formula:

$$A = \pi(3)^2 = 9\pi = 28.3 \text{ ft}^2$$

As the shaded area is outside the circle, the area for the circle can be subtracted from the area of the rectangle to yield an area of 43.7 ft^2.

Geometric figures may be shown as pictures or described in words. If a rectangular playing field with dimensions 95 meters long by 50 meters wide is measured for perimeter, the distance around the field must be found. The perimeter includes two lengths and two widths to measure the entire outside of the field. This quantity can be calculated using the following equation:

$$P = 2(95) + 2(50) = 290\ m$$

The distance around the field is 290 meters.

Perimeter and area are two-dimensional descriptions; volume is three-dimensional. Volume describes the amount of space that an object occupies, but it differs from area because it has three dimensions instead of two. The units for volume are cubic units, such as cubic meters, cubic inches, and cubic miles. Volume can be found by using formulas for common objects such as cylinders and boxes.

The following chart shows a diagram and formula for the volume of two objects.

Shape	**Formula**	**Diagram**
Rectangular Prism (box)	$V = length \times width \times height$	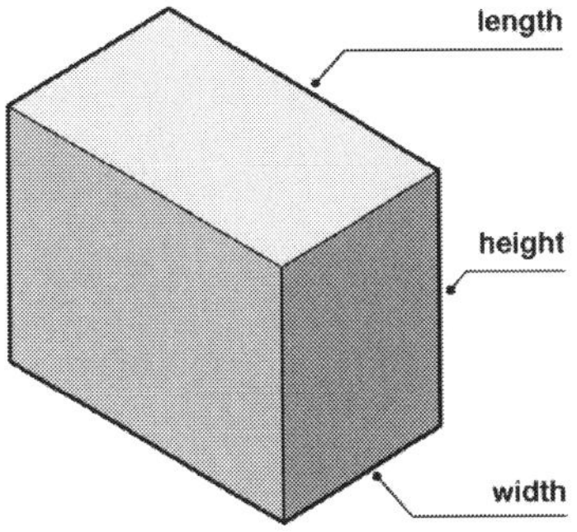
Cylinder	$V = \pi \times radius^2 \times height$	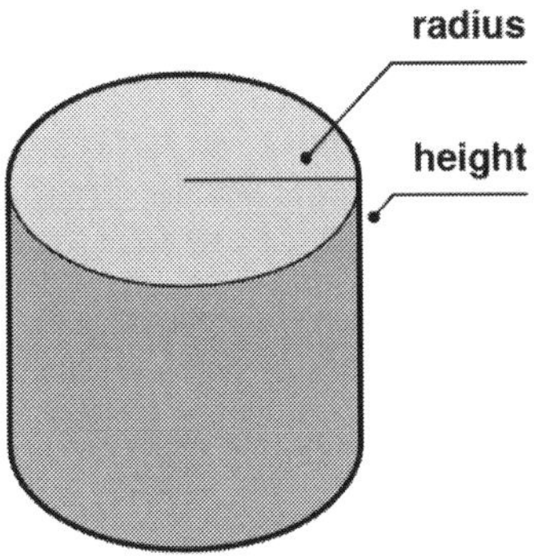

Volume formulas of these two objects are derived by finding the area of the bottom two-dimensional shape, such as the circle or rectangle, and then multiplying times the height of the three-dimensional shape.

Other volume formulas include the volume of a cube with side length s: $V = s^3$; the volume of a sphere with radius r: $V = \frac{4}{3}\pi r^3$; and the volume of a cone with radius r and height h:

$$V = \frac{1}{3}\pi r^2 h$$

If a soda can has a height of 5 inches and a radius on the top of 1.5 inches, the volume can be found using one of the given formulas. A soda can is a cylinder. Knowing the given dimensions, the formula can be completed as follows:

$$V = \pi(radius)^2 \times height$$

$$\pi(1.5 \text{ in})^2 \times 5 \text{ in} = 35.325 \text{ in}^3$$

Notice that the units for volume are inches cubed because it refers to the number of cubic inches required to fill the can.

With any geometric calculations, it's important to determine what dimensions are given and what quantities the problem is asking for. If a connection can be made between them, the answer can be found.

Other geometric quantities can include angles inside a triangle. The sum of the measures of any triangle's three angles is 180 degrees. Therefore, if only two angles are known, the third can be found by subtracting the sum of the two known quantities from 180. Two angles that add up to 90 degrees are known as complementary angles. For example, angles measuring 72 and 18 degrees are complementary. Finally, two angles that add up to 180 degrees are known as supplementary angles. To find the supplement of an angle, subtract the given angle from 180 degrees. For example, the supplement of an angle that is 50 degrees is 180 – 50 = 130 degrees.

These terms involving angles can be seen in many types of word problems. For example, consider the following problem: The measure of an angle is 60 degrees less than two times the measure of its complement. What is the angle's measure? To solve this, let x be the unknown angle. Therefore, its complement is $90 - x$. The problem gives that:

$$x = 2(90 - x) - 60$$

To solve for x, distribute the 2, and collect like terms. This process results in:

$$x = 120 - 2x$$

Then, use the addition property to add $2x$ to both sides to obtain $3x = 120$. Finally, use the multiplication properties of equality to divide both sides by 3 to get $x = 40$. Therefore, the angle measures 40 degrees. Also, its complement measures 50 degrees.

Finding the Volume and Surface Area of Right Rectangular Prisms, Including Those with Fractional Edge Lengths

Right rectangular prisms are those prisms in which all sides are rectangles and all angles are right, or equal to 90 degrees. The volume for these objects can be found by multiplying the length by the width by the height. The formula is $V = lwh$. For the following prism, the volume formula is:

$$V = 6\frac{1}{2} \times 3 \times 9$$

When dealing with fractional edge lengths, it is helpful to convert the length to an improper fraction. The length $6\frac{1}{2}$ cm becomes $\frac{13}{2}$ cm. Then the formula becomes:

$$V = \frac{13}{2} \times 3 \times 9 = \frac{13}{2} \times \frac{3}{1} \times \frac{9}{1} = \frac{351}{2}$$

This value for volume is better understood when turned into a mixed number, which would be $175\frac{1}{2}$ cm^3.

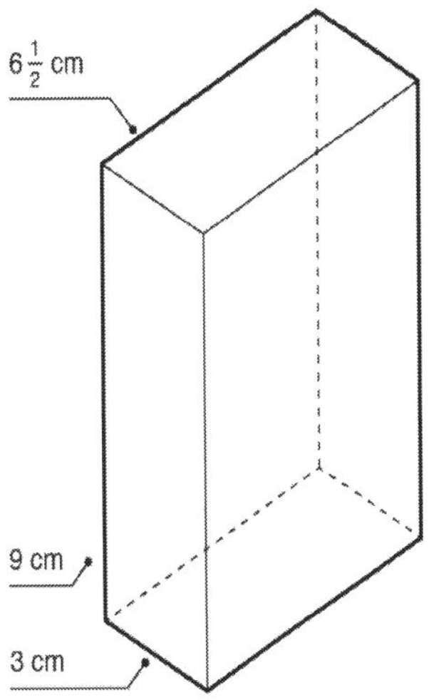

When dimensions for length are given with fractional parts, it can be helpful to turn the mixed number into an improper fraction, then multiply to find the volume, then convert back to a mixed number. When finding surface area, this conversion to improper fractions can also be helpful. The surface area can be found for the same prism above by breaking down the figure into basic shapes. These shapes are rectangles, made up of the two bases, two sides, and the front and back. The formula for surface area adds the areas for each of these shapes in the following equation:

$$SA = 6\frac{1}{2} \times 3 + 6\frac{1}{2} \times 3 + 3 \times 9 + 3 \times 9 + 6\frac{1}{2} \times 9 + 6\frac{1}{2} \times 9$$

Because there are so many terms in a surface area formula and because this formula contains a fraction, it can be simplified by combining groups that are the same. Each set of numbers is used twice, to represent areas for the opposite sides of the prism. The formula can be simplified to:

$$SA = 2\left(6\frac{1}{2} \times 3\right) + 2(3 \times 9) + 2\left(6\frac{1}{2} \times 9\right)$$

$$2\left(\frac{13}{2} \times 3\right) + 2(27) + 2\left(\frac{13}{2} \times 9\right)$$

$$2\left(\frac{39}{2}\right) + 54 + 2\left(\frac{117}{2}\right)$$

$$39 + 54 + 117$$

$$210 \text{ cm}^2$$

Using Trigonometric Ratios and the Pythagorean Theorem to Solve Applied Problems Involving Right Triangles

Within right triangles, trigonometric ratios can be defined for the acute angle within the triangle. Consider the following right triangle. The side across from the right angle is known as the **hypotenuse,** the acute angle being discussed is labeled θ, the side across from the acute angle is known as the **opposite** side, and the other side is known as the **adjacent** side.

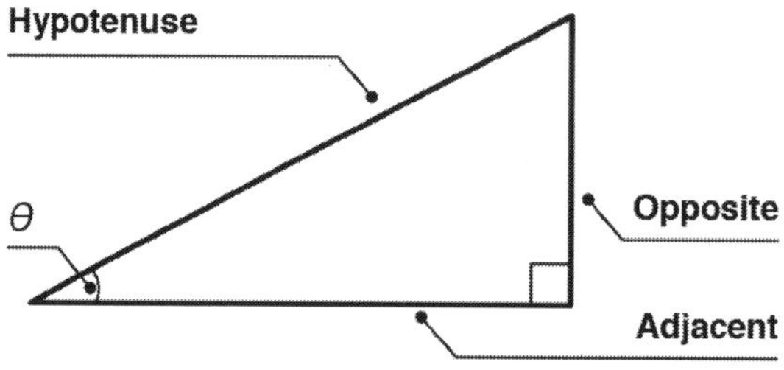

$$\sin\theta = \frac{\text{opposite}}{\text{hypotenuse}} \qquad \csc\theta = \frac{\text{hypotenuse}}{\text{opposite}}$$

$$\cos\theta = \frac{\text{adjacent}}{\text{hypotenuse}} \qquad \sec\theta = \frac{\text{hypotenuse}}{\text{adjacent}}$$

$$\tan\theta = \frac{\text{opposite}}{\text{adjacent}} \qquad \cot\theta = \frac{\text{adjacent}}{\text{opposite}}$$

The six trigonometric ratios are shown above as well. "Sin" is short for sine, "cos" is short for cosine, "tan" is short for tangent, "csc" is short for cosecant, "sec" is short for secant, and "cot" is short for cotangent. A mnemonic device exists that is helpful to remember the ratios. SOHCAHTOA stands for Sine = Opposite/Hypotenuse, Cosine = Adjacent/Hypotenuse, and Tangent = Opposite/Adjacent. The other three trigonometric ratios are reciprocals of sine, cosine, and tangent because $\csc\theta = \frac{1}{\sin\theta}$, $\sec\theta = \frac{1}{\cos\theta}$, and $\cot\theta = \frac{1}{\tan\theta}$.

The **Pythagorean Theorem** expresses an important relationship between the three sides of a right triangle. It states that the square of the hypotenuse is equal to the sum of the squares of the other two sides. When using the Pythagorean Theorem, the hypotenuse is labeled as side c, the opposite is labeled as side a, and the adjacent side is side b.

The theorem can be seen in the following diagram:

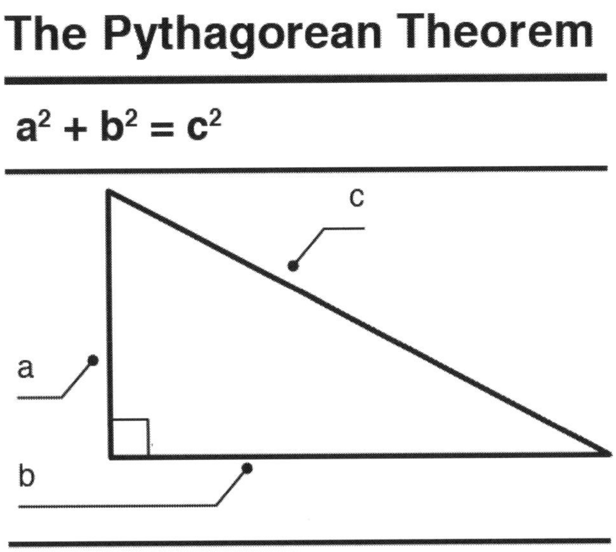

Both the trigonometric ratios and Pythagorean Theorem can be used in problems that involve finding either a missing side or missing angle of a right triangle. Look to see what sides and angles are given and select the correct relationship that will assist in finding the missing value. These relationships can also be used to solve application problems involving right triangles. Often, it is helpful to draw a figure to represent the problem to see what is missing.

Adding, Subtracting, Multiplying, Dividing, and Simplifying Complex Numbers

A **complex number** is of the form $a + bi$, where *a* and *b* are real numbers and *i* is the imaginary number, $\sqrt{-1}$. The real part of the number is *a*, and the imaginary part is *bi*. To add two complex numbers, add the real parts together and the imaginary parts together, similar to combining like terms in polynomials. For instance:

$$(3 + 4i) + (4 - 3i) = 7 + i$$

Subtraction is done in a similar manner, by adding the opposite of the second number.

For example:

$$(3 + 4i) - (4 - 3i) = (3 + 4i) + (-4 + 3i) = -1 + 7i.$$

Multiplication is performed the same way as multiplying polynomials, through foiling and collecting like terms. Remember that $i^2 = -1$.

Therefore:

$$(3 + 4i)(4 - 3i) = 12 - 9i + 16i - 12i^2$$

$$12 + 7i + 12 = 24 + 7i$$

Simplifying complex numbers involves combining like terms: real parts with real parts and imaginary parts with imaginary parts.

The **conjugate** of a complex number $a + bi$ is $a - bi$. Note that when a complex number is multiplied times its conjugate, the result is:

$$(a + bi)(a - bi)$$

$$a^2 - b^2i^2 = a^2 + b^2$$

and its square root is known as the **magnitude,** or **absolute value**, of the number. To divide a complex number by another complex number, multiply the numerator and the denominator of the expression times the conjugate of the denominator. Here is an example of a complex number division problem:

A Complex Number Division Problem

$$\frac{2+3i}{4-5i} = \frac{2+3i}{4-5i} \times \frac{4+5i}{4+5i} =$$

$$= \frac{8 + 10i + 12i + 15i^2}{16 + 20i - 20i - 25i^2} =$$

$$= \frac{8 + 22i + 15 \times (-1)}{16 - 25 \times (-1)} =$$

$$= \frac{-7 + 22i}{49} =$$

$$= -\frac{7}{49} + \frac{22}{49}i$$

Converting Between Degrees and Radians, and Using Radians to Determine Arc Lengths

Angles are measured in both radians and degrees. The symbol for degrees is ° and radians can be abbreviated as rad. A circle is made up of $360°$ or 2π rad. It is true that π radians equal $180°$. If an angle measurement is given in radians, multiply the value times $180°$ then divide by π to obtain the equivalent value in degrees. For instance:

$$\frac{\pi}{2} = \frac{180°\pi}{2\pi} = 90°$$

If an angle measurement is given in degrees, multiply the value times π and then divide by $180°$ to obtain the equivalent value in radians. For instance:

$$45° = \frac{45°\pi}{180°} = \frac{\pi}{4}\ radians$$

Here is a diagram that shows conversions between radians and degrees in a circle:

The Conversions Between Radians and Degrees in a Circle

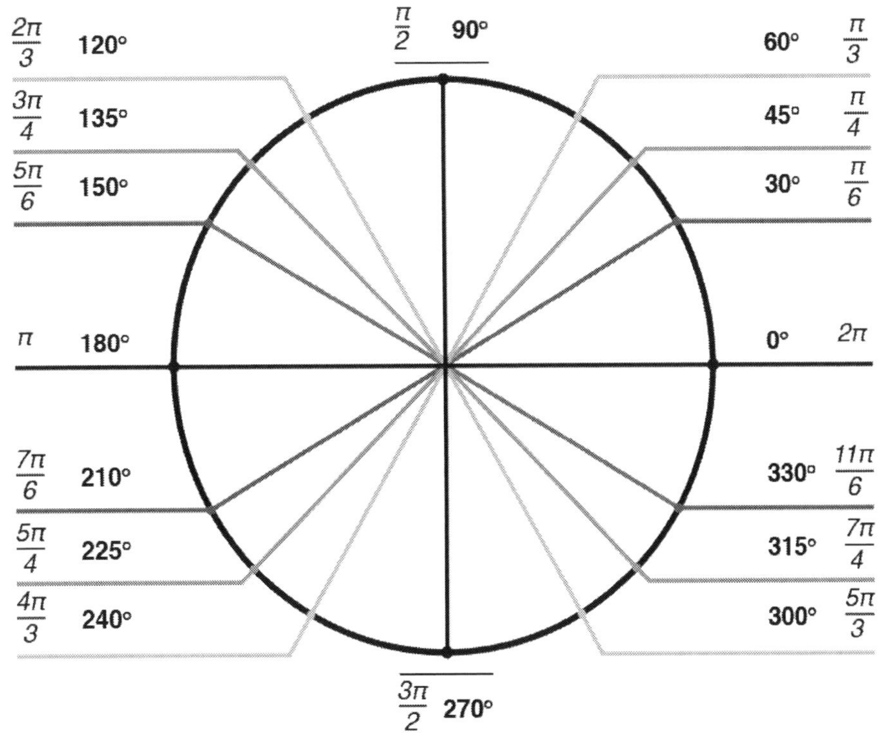

A **circle** is the set of all points the same distance, known as radius *r*, from a single point *C*, known as the center. The **radius** is the length from the center to any point on its edge, and double the radius to obtain the diameter. The **diameter** is the distance of any line reaching two points on the circle that goes through the center. An **arc** on a circle comprises all points between any two points on its edge. A **sector** can be built from an arc and two corresponding radii. Finally, a **central angle** is the angle formed by the intersection of those two radii within a sector.

Here is a diagram that shows the center, radius, central angle, a sector, and an arc of a circle:

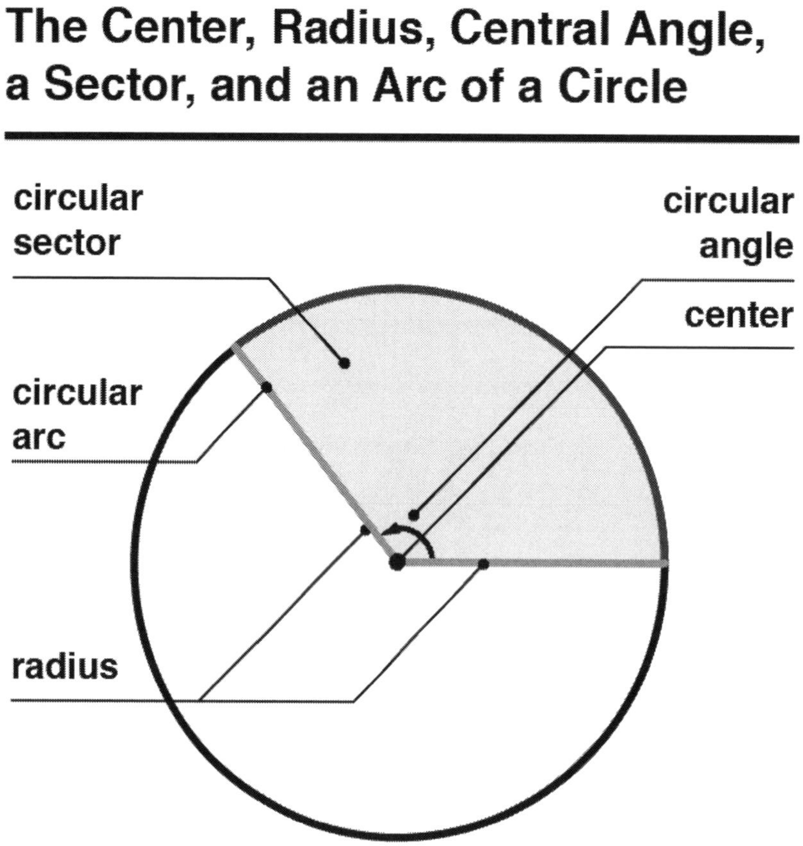

Given an arc, its length is the distance along the curved line that makes up the arc. The formula for the circumference of a circle is $C = 2\pi r$, which is the same as $C = \pi d$. Therefore, the formula for arc length is $2\pi r \frac{A}{360}$, where A is the central angle measurement in degrees, a fraction of the entire circumference. The formula for arc length in radians is rA, where A is the central angle measurement in radians.

Using Trigonometric Functions of Radian Measure

When inputting values into trigonometric functions, the values can either be in degrees or radians. It is important when using calculators to evaluate trigonometric functions that the calculator is in the correct mode: degrees or radians. If a calculator is in the mode in which it is supposed to use degrees and a value in radians is plugged into a trig function, the result will be incorrect.

Another way to evaluate trig functions in degrees and radians is to use the *unit circle*. When using radians, the degree measure used should be the one in radians, which would contain a multiple of π. Here is a unit circle:

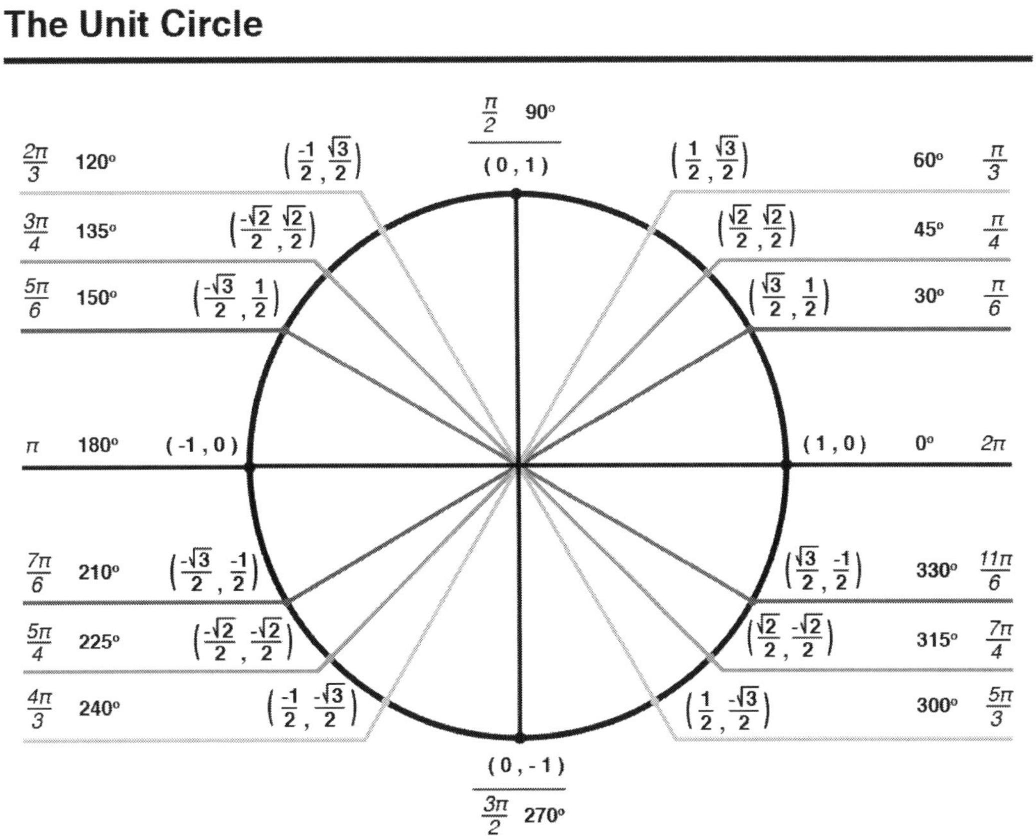

Notice that each angle measure is given in both radians and degrees. The **sine** of an angle is the y-coordinate of the corresponding ordered pair. For instance, $\sin \pi = 0$. The **cosine** of any angle is the x-coordinate of the corresponding ordered pair, and therefore, $\cos \pi = -1$. The **tangent** of any angle is the ratio of the y-coordinate over the x-coordinate, and therefore $\tan \pi = \frac{0}{-1} = 0$. **Cosecant, secant**, and **cotangent** of any angles are the corresponding reciprocals of sine, cosine, and tangent. In this case, $\csc \pi$ is undefined, $\sec \pi = -1$, and $\cot \pi$ is undefined.

Applying Theorems About Circles to Find Arc Lengths, Angle Measures, Chord Lengths, and Areas of Sectors

The formula for area of a circle is $A = \pi r^2$ and therefore, formula for area of a sector is $\pi r^2 \frac{A}{360}$, a fraction of the entire area of the circle. If the radius of a circle and arc length is known, the central angle measurement in degrees can be found by using the formula:

$$\frac{360 \times arclength}{2\pi r}$$

If the desired central angle measurement is in radians, the formula for the central angle measurement is much simpler as:

$$\frac{arc\ length}{r}$$

A **chord** of a circle is a straight-line segment that connects any two points on a circle. The line segment does not have to travel through the center, as the diameter does. Also, note that the chord stops at the circumference of the circle. If it did not stop and extended toward infinity, it would be known as a **secant line.** The following shows a diagram of a circle with a chord shown by the dotted line. The radius is *r* and the central angle is *A*.

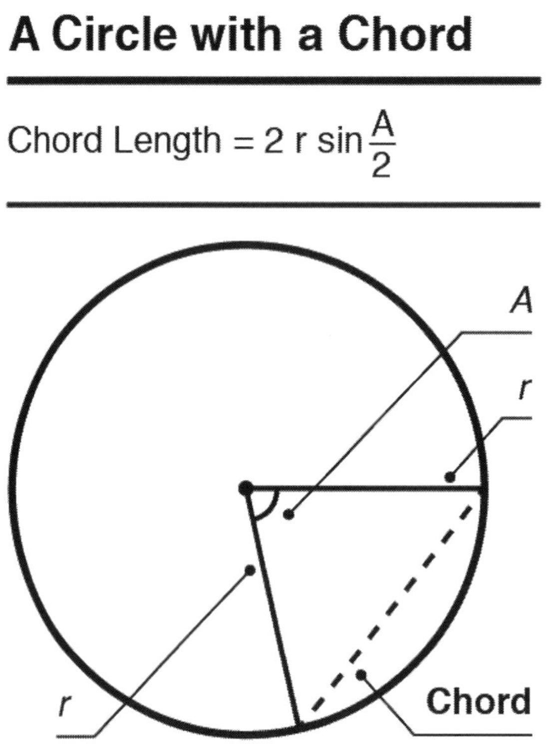

One formula for chord length can be seen in the diagram and is equal to $2r\sin\frac{A}{2}$, where *A* is the central angle. Another formula for chord length is: chord length = $2\sqrt{r^2 - D^2}$, where *D* is equal to the distance from the chord to the center of the circle. This formula is basically a version of the Pythagorean Theorem.

Formulas for chord lengths vary based on what type of information is known. If the radius and central angle are known, the first formula listed above should be used by plugging the radius and angle in directly. If the radius and the distance from the center to the chord are known, the second formula listed previously should be used.

Many theorems exist between arc lengths, angle measures, chord lengths, and areas of sectors. For instance, when two chords intersect in a circle, the product of the lengths of the individual line segments are equal. For instance, in the following diagram, $A \times B = C \times D$.

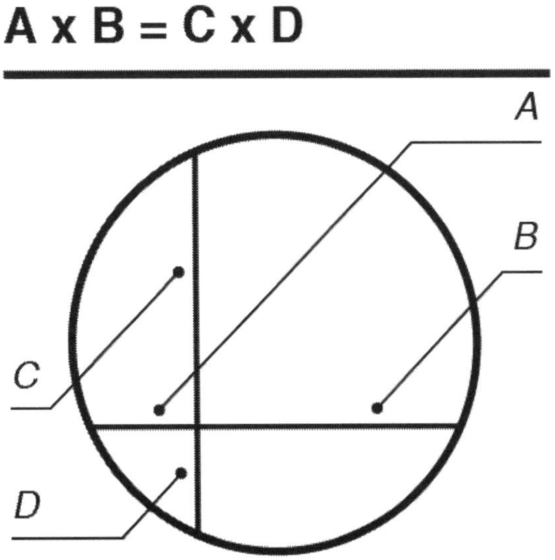

Using Concepts and Theorems about Congruence and Similarity to Solve Problems about Lines, Angles, and Triangles

Two figures are **congruent** if they have the same shape and same size, meaning same angle measurements and equal side lengths. Two figures are similar if they have the same angle measurements but not side lengths. Basically, angles are congruent in similar triangles and their side lengths are constant multiplies of each other. Proving two shapes are similar involves showing that all angles are the same; proving two shapes are congruent involves showing that all angles are the same and that all sides are the same. If two pairs of angles are congruent in two triangles, then those triangles are similar because their third angles have to be equal due to the fact that all three angles add up to 180 degrees.

There are five main theorems that are used to prove congruence in triangles. Each theorem involves showing that different combinations of sides and angles are the same in two triangles, which proves the triangles are congruent. The **side-side-side (SSS)** theorem states that if all sides are equal in two triangles, the triangles are congruent. The **side-angle-side (SAS)** theorem states that if two pairs of sides and the included angles are equal in two triangles then the triangles are congruent. Similarly, the **angle-side-angle (ASA)** theorem states that if two pairs of angles and the included side lengths are equal in two triangles, the triangles are similar. The **angle-angle-side (AAS) theorem** states that two triangles are congruent if they have two pairs of congruent angles and a pair of corresponding equal side lengths that are not included. Finally, the **hypotenuse-leg (HL)** theorem states that if two right triangles have equal hypotenuses and an equal pair of shorter sides, the triangles are congruent. An important item to note is that angle-angle-angle (AAA) is not enough information to prove congruence because the three angles could be equal in two triangles, but their sides could be different lengths.

Using the Relationship Between Similarity, Right Triangles, and Trigonometric Ratios

Within two similar triangles, corresponding side lengths are proportional, and angles are equal. In other words, regarding corresponding sides in two similar triangles, the ratio of side lengths is the same. Recall that the SAS theorem for similarity states that if an angle in one triangle is congruent to an angle in a second triangle, and the lengths of the sides in both triangles are proportional, then the triangles are similar. Also, because the ratio of two sides in two similar right triangles is the same, the trigonometric ratios in similar right triangles are always going to be equal.

If two triangles are similar, and one is a right triangle, the other is a right triangle. The definition of similarity ensures that each triangle has a 90-degree angle. In a similar sense, if two triangles are right triangles containing a pair of equal acute angles, the triangles are similar because the third pair of angles must be equal as well. However, right triangles are not necessarily always similar.

The following triangles are similar:

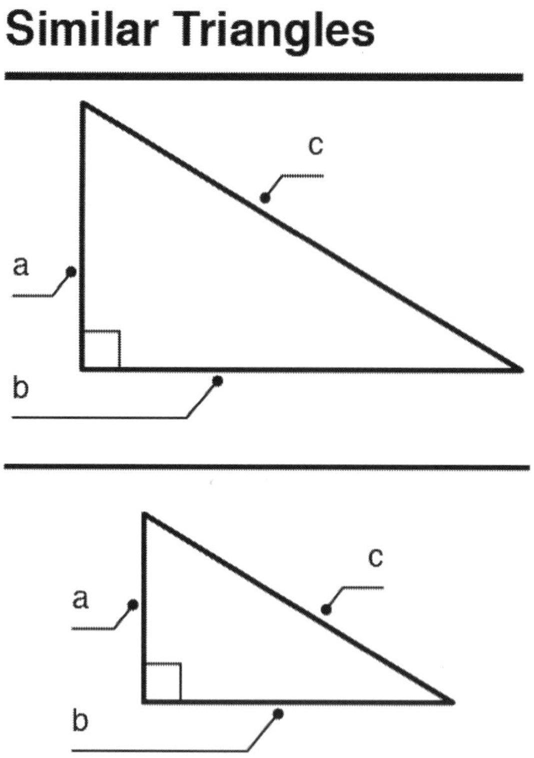

This similarity is not apparent at first glance; however, theorems can be used to show similarity. The Pythagorean Theorem can be used to find the missing side lengths in both triangles. In the larger triangle, the missing side is the hypotenuse, *c*. Therefore:

$$9^2 + 12^2 = c^2$$

This equation is equivalent to $225 = c^2$, so taking the square root of both sides results in the positive root $c = 15$. In the other triangle, the Pythagorean theorem can be used to find the missing side length *b*. The theorem shows that $6^2 + b^2 = 10^2$, and b is then solved for to obtain $b = 8$. The ratio of the sides in the larger triangle to the sides in the smaller triangle is the same value, 1.5. Therefore, the sides are proportional. Because they are both right triangles, they have a congruent angle. The SAS theorem for similarity can be used to show that these two triangles are similar.

Using the Relationship Between Sine and Cosine of Complementary Angles

Two **complementary angles** add up to 90 degrees, or π radians. Within a right triangle, the sine of an angle is equal to the ratio of the side opposite the angle to the hypotenuse and the cosine of an angle is equal to the ratio of the side adjacent to the angle to the hypotenuse. Within a right triangle, there is a right angle and because the sum of all angles in a triangle is 180 degrees, the two other angles add up to 90 degrees, and are therefore complementary. Consider the following right triangle with angles A, B, and C, and sides a, b, and c.

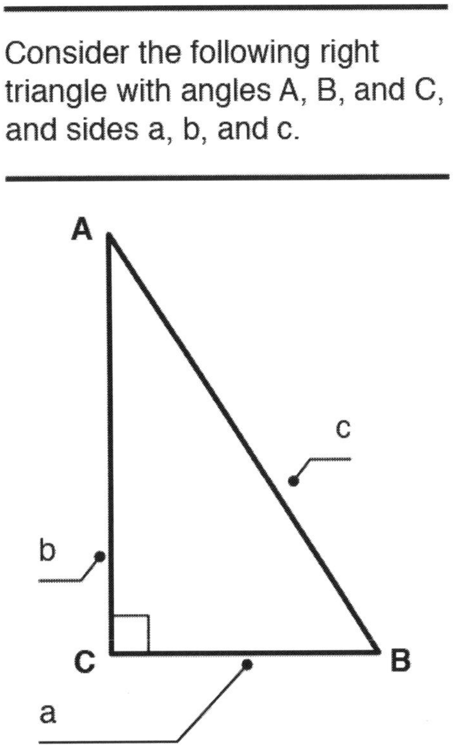

It is true by using such ratios described above that $\sin B = \frac{b}{c}$ and $\cos B = \frac{a}{c}$. Also, it is true that $\cos A = \frac{b}{c}$ and $\sin A = \frac{a}{c}$. Therefore, $\sin B = \cos A$ and $\cos B = \sin A$. A and B are complementary angles, so given two complementary angles, the sine of one equals the cosine of the other, and the cosine of one equals the sine of the other. Given the two complementary angles 30 degrees and 60 degrees, $\sin 30 = \frac{1}{2}$, $\cos 60 = \frac{1}{2}$, $\cos 30 = \frac{\sqrt{3}}{2}$, and $\sin 60 = \frac{\sqrt{3}}{2}$. Either a calculator set in degrees mode, a unit circle, or the Pythagorean Theorem could be used to find all of these values.

Creating or Using an Equation in Two Variables to Solve a Problem About a Circle in the Coordinate Plane

Recall that a circle is the set of all points the same distance, known as radius r, from a single point C, known as the center of the circle. The center has coordinates (h, k) and any point on the circle is an ordered pair with coordinates (x, y). A right triangle with hypotenuse r can be formed with these two points as seen here:

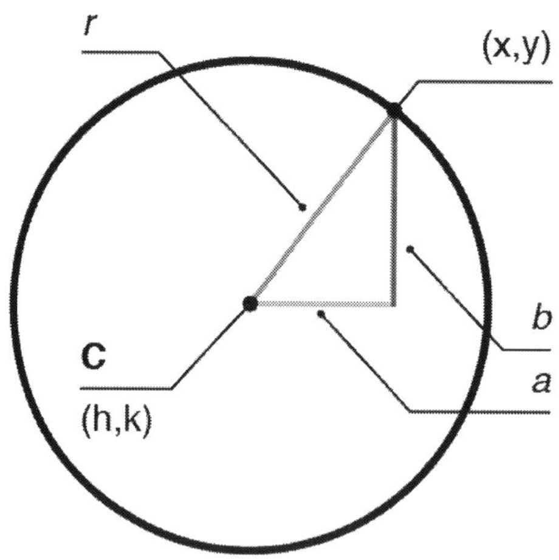

The other side lengths are a and b. The Pythagorean Theorem states that:

$$a^2 + b^2 = r^2$$

However, a can be replaced by $|x - h|$ and b can be replaced by $|y - k|$ because the distance between any two coordinates in the coordinate plane is the absolute value of their difference. That substitution gives:

$$(x - h)^2 + (y - k)^2 = r^2$$

which is the formula used to find the equation of any circle with center (h, k) and radius r. Therefore, if any problem gives the coordinates of the center of a circle and its radius length, this is the equation in two variables that allows any other point on the circle to be found.

Oftentimes, the center or the radius of a circle are not easily seen in the given equation of the circle. If the equation is in standard form of a polynomial equation like:

$$ax^2 + ay^2 + cx + dy + e = 0$$

the algebraic technique of completing the square must be used to find the coordinates of the center and the radius. Completing the square must be done within both variables x and y. First, the constant term needs to be subtracted off of both sides of the equation, and then the x and y terms need to be grouped together. Then, the entire equation needs to be divided by a. Then, divide the coefficient of the x term by 2, square it, and add that value to both sides of the equation. This value should be grouped with the x terms. Next, divide the coefficient of the y term by 2, square it, and add it to both sides of the equation, grouping it with the y terms. The trinomial in both x and y are now perfect square trinomials and can be factored into squares of a binomial. This process results in:

$$(x - h)^2 + (y - k)^2 = r^2$$

showing the radius and coordinates of the center.

Practice Questions

1. Given that $abc \neq 0$, which of the following is equivalent to the expression $\frac{a^5b^3c^7}{(ab^2c^3)^2}$?

a. a^3bc
b. a^3b^2c
c. $\frac{a^3c}{b}$
d. $\frac{ac}{b}$

2. A shoe salesclerk earns $8x + 0.00048y^2$ dollars, where x is the total number of hours worked and y represents her total sales in dollars. If she earned \$200 dollars and had \$500 in sales, how many hours did she work?

a. 8
b. 10
c. 12
d. 14

3. If $x < 0$ and $6x^2 - 14x - 20 = 0$, then what is x?

a. -1
b. $-\frac{10}{3}$
c. -20
d. $-\frac{1}{3}$

4. A system of linear equations in two variables has no solutions. One of the lines is drawn below. Which of the following equations could the other line be?

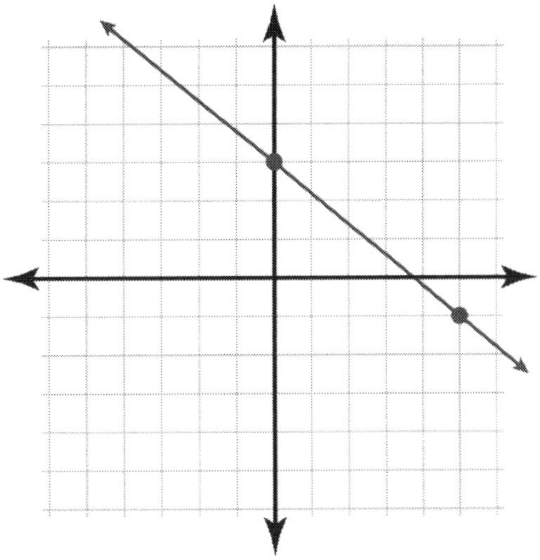

a. $y = \frac{5}{4}x + 4$
b. $y = -\frac{4}{5}x + 16$
c. $y = 5x + 3$
d. $y = -\frac{5}{4}x + 3$

5. What are three consecutive odd integers that satisfy the following: the sum of four times the first, two times the second, and two less than the third is equal to 111?

a. 9, 11, 13
b. 11, 13, 15
c. 13, 15, 19
d. 15, 17, 19

Answer Explanations

1. C: First, the parentheses are removed to obtain:

$$\frac{a^5b^3c^7}{a^2b^4c^6}$$

Then, applying exponential law $\frac{x^m}{x^n} = x^{m-n}$ simplifies the expression into:

$$\frac{a^3c}{b}$$

2. B: First, plug 500 in for *y*, which represents her sales. Then, set this expression equal to 200, her total earnings, for:

$$8x + 0.00048(500)^2 = 200$$

Simplifying:

$$8x + 120 = 200$$

Solving this for *x* gives x = 10, and therefore, she worked 10 hours.

3. A: First, factor the polynomial completely. It factors into:

$$2(3x^2 - 7x - 10)$$

$$2(3x - 10)(x + 1) = 0$$

Then, set each factor equal to 0 and solve for *x* to obtain the two solutions $\frac{10}{3}$ and -1. The problem specifies that we are looking for a negative solution, so the answer is -1.

4. B: For a system of equations to have no solution, the equations must be parallel. Parallel lines have the same slope and different *y*-intercepts. The equation of the line shown is:

$$y = -\frac{4}{5}x + 3$$

Therefore, the only line given that is parallel is $y = -\frac{4}{5}x + 16$ because it has the same slope and different *y*-intercepts.

5. D: Let *x* represent the first odd integer. The next odd integer would then be *x* + 2, and the third would be *x* + 4. The problem states that four times the first (4*x*) plus 2 times the second (2(*x* + 2)) plus 2 less than the third (*x* + 4 - 2) add up to 111. Therefore:

$$4x + 2(x + 2) + x + 4 - 2 = 111$$

The left side simplifies to $7x + 6$. The solution of $7x + 6 = 111$ is $x = 15$. Therefore, the three consecutive odd integers are 15, 17, and 19.

SAT Essay

Using Grammar to Enhance Clarity in Writing

Possessives

Possessive forms indicate possession, i.e. that something belongs to or is owned by someone or something. As such, the most common parts of speech to be used in possessive form are adjectives, nouns, and pronouns. The rule for correctly spelling/punctuating possessive nouns and proper nouns is with *-'s,* like "the woman's briefcase" or "Frank's hat." With possessive adjectives, however, apostrophes are not used: these include *my, your, his, her, its, our,* and *their,* like "my book," "your friend," "his car," "her house," "its contents," "our family," or "their property." Possessive pronouns include *mine, yours, his, hers, its, ours,* and *theirs.* These also have no apostrophes. The difference is that possessive adjectives take direct objects, whereas possessive pronouns replace them. For example, instead of using two possessive adjectives in a row, as in "I forgot my book, so Blanca let me use her book," which reads monotonously, replacing the second one with a possessive pronoun reads better: "I forgot my book, so Blanca let me use hers."

Pronouns

There are three pronoun cases: subjective case, objective case, and possessive case. Pronouns as subjects are pronouns that replace the subject of the sentence, such as *I, you, he, she, it, we, they* and *who.* Pronouns as objects replace the object of the sentence, such as *me, you, him, her, it, us, them,* and *whom.* Pronouns that show possession are *mine, yours, hers, its, ours, theirs,* and *whose.* The following are examples of different pronoun cases:

- Subject pronoun: *She* ate the cake for her birthday. *I* saw the movie.
- Object pronoun: You gave *me* the card last weekend. She gave the picture to *him.*
- Possessive pronoun: That bracelet you found yesterday is *mine. His* name was Casey.

Adjectives

Adjectives are descriptive words that modify nouns or pronouns. They may occur before or after the nouns or pronouns they modify in sentences. For example, in "This is a big house," *big* is an adjective modifying or describing the noun *house.* In "This house is big," the adjective is at the end of the sentence rather than preceding the noun it modifies.

A rule of punctuation that applies to adjectives is to separate a series of adjectives with commas. For example, "Their home was a large, rambling, old, white, two-story house." A comma should never separate the last adjective from the noun, though.

Adverbs

Whereas adjectives modify and describe nouns or pronouns, **adverbs** modify and describe adjectives, verbs, or other adverbs. Adverbs can be thought of as answers to questions in that they describe when, where, how, how often, how much, or to what extent.

Many (but not all) adjectives can be converted to adverbs by adding *–ly.* For example, in "She is a quick learner," *quick* is an adjective modifying *learner.* In "She learns quickly," *quickly* is an adverb modifying *learns.* One exception is *fast. Fast* is an adjective in "She is a fast learner." However, *–ly* is never added to the word *fast,* it retains the same form as an adverb in "She learns fast."

Verbs

A **verb** is a word or phrase that expresses action, feeling, or state of being. Verbs explain what their subject is *doing*. Three different types of verbs used in a sentence are action verbs, linking verbs, and helping verbs.

Action verbs show a physical or mental action. Some examples of action verbs are *play, type, jump, write, examine, study, invent, develop,* and *taste*. The following example uses an action verb:

> Kat *imagines* that she is a mermaid in the ocean.

The verb *imagines* explains what Kat is doing: she is imagining being a mermaid.

Linking verbs connect the subject to the predicate without expressing an action. The following sentence shows an example of a linking verb:

> The mango *tastes* sweet.

The verb *tastes* is a linking verb. The mango doesn't *do* the tasting, but the word *taste* links the mango to its predicate, sweet. Most linking verbs can also be used as action verbs, such as *smell, taste, look, seem, grow,* and *sound*. Saying something *is* something else is also an example of a linking verb. For example, if we were to say, "Peaches is a dog," the verb *is* would be a linking verb in this sentence, since it links the subject to its predicate.

Helping verbs are verbs that help the main verb in a sentence. Examples of helping verbs are *be, am, is, was, have, has, do, did, can, could, may, might, should,* and *must,* among others. The following are examples of helping verbs:

> Jessica *is* planning a trip to Hawaii.
>
> Brenda *does* not like camping.
>
> Xavier *should* go to the dance tonight.

Notice that after each of these helping verbs is the main verb of the sentence: *planning, like,* and *go*. Helping verbs usually show an aspect of time.

Transitional Words and Phrases

In writing, some sentences naturally lead to others, whereas in other cases, a new sentence expresses a new idea. Transitional phrases connect sentences and the ideas they convey, which makes the writing coherent. Transitional language also guides the reader from one thought to the next. For example, when pointing out an objection to the previous idea, starting a sentence with "However," "But," or "On the other hand" is transitional. When adding another idea or detail, writers use "Also," "In addition," "Furthermore," "Further," "Moreover," "Not only," etc. Readers have difficulty perceiving connections between ideas without such transitional wording.

Subject-Verb Agreement

Lack of subject-verb agreement is a very common grammatical error. One of the most common instances is when people use a series of nouns as a compound subject with a singular instead of a plural verb. Here is an example:

> Identifying the best books, locating the sellers with the lowest prices, and paying for them *is* difficult

instead of saying "*are* difficult." Additionally, when a sentence subject is compound, the verb is plural:

He and his cousins *were* at the reunion.

However, if the conjunction connecting two or more singular nouns or pronouns is "or" or "nor," the verb must be singular to agree:

That pen or another one like it is in the desk drawer.

If a compound subject includes both a singular noun and a plural one, and they are connected by "or" or "nor," the verb must agree with the subject closest to the verb: "Sally or her sisters go jogging daily"; but "Her sisters or Sally goes jogging daily."

Simply put, singular subjects require singular verbs and plural subjects require plural verbs. A common source of agreement errors is not identifying the sentence subject correctly. For example, people often write sentences incorrectly like, "The group of students *were* complaining about the test." The subject is not the plural "students" but the singular "group." Therefore, the correct sentence should read, "The group of students *was* complaining about the test." The converse also applies, for example, in this incorrect sentence: "The facts in that complicated court case *is* open to question." The subject of the sentence is not the singular "case" but the plural "facts." Hence the sentence would correctly be written: "The facts in that complicated court case *are* open to question." New writers should not be misled by the distance between the subject and verb, especially when another noun with a different number intervenes as in these examples. The verb must agree with the subject, not the noun closest to it.

Pronoun-Antecedent Agreement

Pronouns within a sentence must refer specifically to one noun, known as the **antecedent.** Sometimes, if there are multiple nouns within a sentence, it may be difficult to ascertain which noun belongs to the pronoun. It's important that the pronouns always clearly reference the nouns in the sentence so as not to confuse the reader. Here's an example of an unclear pronoun reference:

After Catherine cut Libby's hair, David bought her some lunch.

The pronoun in the examples above is *her.* The pronoun could either be referring to *Catherine* or *Libby.* Here are some ways to write the above sentence with a clear pronoun reference:

After Catherine cut Libby's hair, David bought Libby some lunch.

David bought Libby some lunch after Catherine cut Libby's hair.

But many times the pronoun will clearly refer to its antecedent, like the following:

After David cut Catherine's hair, he bought her some lunch.

Formal and Informal Language

Formal language is less personal than informal language. It is more "buttoned-up" and business-like, adhering to proper grammatical rules. It is used in professional or academic contexts, to convey respect or authority. For example, one would use formal language to write an informative or argumentative essay for school or to address a superior. Formal language avoids contractions, slang, colloquialisms, and first-person pronouns. Formal language uses sentences that are usually more complex and often in passive voice. Punctuation can differ as well. For example, exclamation points (!) are used to show strong emotion or can be used as an interjection but should be used sparingly in formal writing situations.

Informal language is often used when communicating with family members, friends, peers, and those known more personally. It is more casual, spontaneous, and forgiving in its conformity to grammatical rules and conventions. Informal language is used for personal emails and correspondence between coworkers or other familial relationships. The tone is more relaxed. In informal writing, slang, contractions, clichés, and the first- and second-person are often used.

Developing a Well-Organized Paragraph

A paragraph is a series of connected and related sentences addressing one topic. Writing good paragraphs benefits writers by helping them to stay on target while drafting and revising their work. It benefits readers by helping them to follow the writing more easily. Regardless of how brilliant their ideas may be, writers who do not present them in organized ways will fail to engage readers—and fail to accomplish their writing goals. A fundamental rule for paragraphing is to confine each paragraph to a single idea. When writers find themselves transitioning to a new idea, they should start a new paragraph. However, a paragraph can include several pieces of evidence supporting its single idea; and it can include several points if they are all related to the overall paragraph topic. When writers find each point becoming lengthy, they may choose instead to devote a separate paragraph to every point and elaborate upon each more fully.

An effective paragraph should have these elements:

- Unity: One major discussion point or focus should occupy the whole paragraph from beginning to end.
- Coherence: For readers to understand a paragraph, it must be coherent. Two components of coherence are logical and verbal bridges. In logical bridges, the writer may write consecutive sentences with parallel structure or carry an idea over across sentences. In verbal bridges, writers may repeat key words across sentences.
- A topic sentence: The paragraph should have a sentence that generally identifies the paragraph's thesis or main idea.
- Sufficient development: To develop a paragraph, writers can use the following techniques after stating their topic sentence:
    - Define terms
    - Cite data
    - Use illustrations, anecdotes, and examples
    - Evaluate causes and effects
    - Analyze the topic
    - Explain the topic using chronological order

A **topic sentence** identifies the main idea of the paragraph. Some are explicit, some implicit. The topic sentence can appear anywhere in the paragraph. However, many experts advise beginning writers to place each paragraph topic sentence at or near the beginning of its paragraph to ensure that their readers understand what the topic of each paragraph is. Even without having written an explicit topic sentence, the writer should still be able to summarize readily what subject matter each paragraph addresses. The writer must then fully develop the topic that is introduced or identified in the topic sentence. Depending on what the writer's purpose is, they may use different methods for developing each paragraph.

Two main steps in the process of organizing paragraphs and essays should both be completed after determining the writing's main point, while the writer is planning or outlining the work. The initial step is to give an order to the topics addressed in each paragraph. Writers must have logical reasons for putting one paragraph first, another second, etc. The second step is to sequence the sentences in each paragraph. As with the first step, writers must have logical reasons for the order of sentences. Sometimes the work's main point obviously indicates a specific order.

Topic Sentences

To be effective, a topic sentence should be concise so that readers get its point without losing the meaning among too many words. As an example, in *Only Yesterday: An Informal History of the 1920s* (1931), author Frederick Lewis Allen's topic sentence introduces his paragraph describing the 1929 stock market crash: "The Bull Market was dead." This example illustrates the criteria of conciseness and brevity. It is also a strong sentence, expressed clearly and unambiguously. The topic sentence also introduces the paragraph, alerting the reader's attention to the main idea of the paragraph and the subject matter that follows the topic sentence.

Experts often recommend opening a paragraph with the topic sentences to enable the reader to realize the main point of the paragraph immediately. Application letters for jobs and university admissions also benefit from opening with topic sentences. However, positioning the topic sentence at the end of a paragraph is more logical when the paragraph identifies a number of specific details that accumulate evidence and then culminates with a generalization. While paragraphs with extremely obvious main ideas need no topic sentences, more often—and particularly for students learning to write—the topic sentence is the most important sentence in the paragraph. It not only communicates the main idea quickly to readers; it also helps writers produce and control information.

Practice Test #1

Reading

The next six questions are based on the following passage from The Life, Crime, and Capture of John Wilkes Booth by George Alfred Townsend.

The box in which the President sat consisted of two boxes turned into one, the middle partition being removed, as on all occasions when a state party visited the theater. The box was on a level with the dress circle; about twelve feet above the stage. There were two entrances—the door nearest to the wall having been closed and locked; the door nearest the balustrades of the dress circle, and at right angles with it, being open and left open, after the visitors had entered. The interior was carpeted, lined with crimson paper, and furnished with a sofa covered with crimson velvet, three arm chairs similarly covered, and six cane-bottomed chairs. Festoons of flags hung before the front of the box against a background of lace.

President Lincoln took one of the arm-chairs and seated himself in the front of the box, in the angle nearest the audience, where, partially screened from observation, he had the best view of what was transpiring on the stage. Mrs. Lincoln sat next to him, and Miss Harris in the opposite angle nearest the stage. Major Rathbone sat just behind Mrs. Lincoln and Miss Harris. These four were the only persons in the box.

The play proceeded, although "Our American Cousin," without Mr. Sothern, has, since that gentleman's departure from this country, been justly esteemed a very dull affair. The audience at Ford's, including Mrs. Lincoln, seemed to enjoy it very much. The worthy wife of the President leaned forward, her hand upon her husband's knee, watching every scene in the drama with amused attention. Even across the President's face at intervals swept a smile, robbing it of its habitual sadness.

About the beginning of the second act, the mare, standing in the stable in the rear of the theater, was disturbed in the midst of her meal by the entrance of the young man who had quitted her in the afternoon. It is presumed that she was saddled and bridled with exquisite care.

Having completed these preparations, Mr. Booth entered the theater by the stage door; summoned one of the scene shifters, Mr. John Spangler, emerged through the same door with that individual, leaving the door open, and left the mare in his hands to be held until he (Booth) should return. Booth who was even more fashionably and richly dressed than usual, walked thence around to the front of the theater, and went in. Ascending to the dress circle, he stood for a little time gazing around upon the audience and occasionally upon the stage in his usual graceful manner. He was subsequently observed by Mr. Ford, the proprietor of the theater, to be slowly elbowing his way through the crowd that packed the rear of the dress circle toward the right side, at the extremity of which was the box where Mr. and Mrs. Lincoln and their companions were seated. Mr. Ford casually noticed this as a slightly extraordinary symptom of interest on the part of an actor so familiar with the routine of the theater and the play.

1. Which of the following best describes the author's attitude toward the events leading up to the assassination of President Lincoln?

a. Excitement due to the setting and its people.
b. Sadness due to the death of a beloved president.
c. Anger because of the impending violence.
d. Neutrality due to the style of the report.

2. What does the author mean by the last sentence in the passage?

a. Mr. Ford was suspicious of Booth and assumed he was making his way to Mr. Lincoln's box.
b. Mr. Ford assumed Booth's movement throughout the theater was due to being familiar with the theater.
c. Mr. Ford thought that Booth was making his way to the theater lounge to find his companions.
d. Mr. Ford thought that Booth was elbowing his way to the dressing room to get ready for the play.

3. Given the author's description of the play "Our American Cousin," which one of the following is most analogous to Mr. Sothern's departure from the theater?

a. A ballet dancer who leaves the New York City Ballet just before they go on to their final performance.
b. A basketball player leaves an NBA team and the next year they make it to the championship but lose.
c. A lead singer leaves their band to begin a solo career, and the band drops in sales by 50 percent on their next album.
d. A movie actor who dies in the middle of making a movie and the movie is made anyway by actors who resemble the deceased.

4. Based on the organizational structure of the passage, which of the following texts most closely relates?

a. A chronological account in a fiction novel of a woman and a man meeting for the first time.
b. A cause-and-effect text ruminating on the causes of global warming.
c. An autobiography that begins with the subject's death and culminates to his birth.
d. A text focusing on finding a solution to the problem of the Higgs boson particle.

5. Which of the following words, if substituted for the word *festoons* in the first paragraph, would LEAST change the meaning of the sentence?

a. Feathers
b. Armies
c. Adornments
d. Buckets

6. What is the primary purpose of the passage?

a. To persuade the audience that John Wilkes Booth killed Abraham Lincoln.
b. To inform the audience of the setting wherein Lincoln was shot.
c. To narrate the bravery of Lincoln and his last days as President.
d. To recount in detail the events that led up to Abraham Lincoln's death.

The next seven questions are based on the following passage from The Story of Germ Life by Herbert William Conn.

The first and most universal change effected in milk is its souring. So universal is this phenomenon that it is generally regarded as an inevitable change which can not be avoided, and, as already pointed out, has in the past been regarded as a normal property of milk. To-day, however, the phenomenon is well understood. It is due to the action of certain of the milk bacteria upon the milk sugar which converts it into lactic acid, and this acid gives the sour taste and curdles the milk. After this acid is produced in small quantity its presence proves deleterious to the growth of the bacteria, and further bacterial growth is checked. After souring, therefore, the milk for some time does not ordinarily undergo any further changes.

Milk souring has been commonly regarded as a single phenomenon, alike in all cases. When it was first studied by bacteriologists it was thought to be due in all cases to a single species of micro-organism which was discovered to be commonly present and named Bacillus acidi lactici. This bacterium has certainly the power of souring milk rapidly, and is found to be very common in dairies in Europe. As soon as bacteriologists turned their attention more closely to the subject it was found that the spontaneous souring of milk was not always caused by the same species of bacterium. Instead of finding this Bacillus acidi lactici always present, they found that quite a number of different species of bacteria have the power of souring milk, and are found in different specimens of soured milk. The number of species of bacteria which have been found to sour milk has increased until something over a hundred are known to have this power. These different species do not affect the milk in the same way. All produce some acid, but they differ in the kind and the amount of acid, and especially in the other changes which are affected at the same time that the milk is soured, so that the resulting soured milk is quite variable. In spite of this variety, however, the most recent work tends to show that the majority of cases of spontaneous souring of milk are produced by bacteria which, though somewhat variable, probably constitute a single species, and are identical with the Bacillus acidi lactici. This species, found common in the dairies of Europe, according to recent investigations occurs in this country as well. We may say, then, that while there are many species of bacteria infesting the dairy which can sour the milk, there is one which is more common and more universally found than others, and this is the ordinary cause of milk souring.

When we study more carefully the effect upon the milk of the different species of bacteria found in the dairy, we find that there is a great variety of changes which they produce when they are allowed to grow in milk. The dairyman experiences many troubles with his milk. It sometimes curdles without becoming acid. Sometimes it becomes bitter, or acquires an unpleasant "tainted" taste, or, again, a "soapy" taste. Occasionally a dairyman finds his milk becoming slimy, instead of souring and curdling in the normal fashion. At such times, after a number of hours, the milk becomes so slimy that it can be drawn into long threads. Such an infection proves very troublesome, for many a time it persists in spite of all attempts made to remedy it. Again, in other cases the milk will turn blue, acquiring about the time it becomes sour a beautiful sky-blue colour. Or it may become red, or occasionally yellow. All of these troubles the dairyman owes to the presence in his milk of unusual species of bacteria which grow there abundantly.

7. The word *deleterious* in the first paragraph can be best interpreted as referring to which one of the following?

a. Amicable
b. Smoldering
c. Luminous
d. Ruinous

8. Which of the following best explains how the passage is organized?

a. The author begins by presenting the effects of a phenomenon, then explains the process of this phenomenon, and then ends by giving the history of the study of this phenomenon.
b. The author begins by explaining a process or phenomenon, then gives the history of the study of this phenomenon, then ends by presenting the effects of this phenomenon.
c. The author begins by giving the history of the study of a certain phenomenon, then explains the process of this phenomenon, then ends by presenting the effects of this phenomenon.
d. The author begins by giving a broad definition of a subject, then presents more specific cases of the subject, then ends by contrasting two different viewpoints on the subject.

9. What is the primary purpose of the passage?

a. To inform the reader of the phenomenon, investigation, and consequences of milk souring.
b. To persuade the reader that milk souring is due to Bacillus acidi lactici, found commonly in the dairies of Europe.
c. To describe the accounts and findings of researchers studying the phenomenon of milk souring.
d. To discount the former researchers' opinions on milk souring and bring light to new investigations.

10. What does the author say about the ordinary cause of milk souring?

a. Milk souring is caused mostly by a species of bacteria called Bacillus acidi lactici, although former research asserted that it was caused by a variety of bacteria.
b. The ordinary cause of milk souring is unknown to current researchers, although former researchers thought it was due to a species of bacteria called Bacillus acidi lactici.
c. Milk souring is caused mostly by a species of bacteria identical to that of Bacillus acidi lactici, although there are a variety of other bacteria that cause milk souring as well.
d. The ordinary cause of milk souring will sometimes curdle without becoming acidic, though sometimes it will turn colors other than white, or have strange smells or tastes.

11. The author of the passage would most likely agree most with which of the following?

a. Milk researchers in the past have been incompetent and have sent us on a wild goose chase when determining what causes milk souring.
b. Dairymen are considered more expert in the field of milk souring than milk researchers.
c. The study of milk souring has improved throughout the years, as we now understand more of what causes milk souring and what happens afterward.
d. Any type of bacteria will turn milk sour, so it's best to keep milk in an airtight container while it is being used.

12. Given the author's account of the consequences of milk souring, which of the following is most closely analogous to the author's description of what happens after milk becomes slimy?

a. The chemical change that occurs when a firework explodes.
b. A rainstorm that overwaters a succulent plant.
c. Mercury inside of a thermometer that leaks out.
d. A child who swallows flea medication.

13. What type of paragraph would most likely come after the third?
   a. A paragraph depicting the general effects of bacteria on milk.
   b. A paragraph explaining a broad history of what researchers have found in regard to milk souring.
   c. A paragraph outlining the properties of milk souring and the way in which it occurs.
   d. A paragraph showing the ways bacteria infiltrate milk and ways to avoid this infiltration.

Passage A

(from "Free Speech in War Time" by James Parker Hall, written in 1921, published in Columbia Law Review, Vol. 21 No. 6)

In approaching this problem of interpretation, we may first put out of consideration certain obvious limitations upon the generality of all guaranties of free speech. An occasional unthinking malcontent may urge that the only meaning not fraught with danger to liberty is the literal one that no utterance may be forbidden, no matter what its intent or result; but in fact it is nowhere seriously argued by anyone whose opinion is entitled to respect that direct and intentional incitations to crime may not be forbidden by the state. If a state may properly forbid murder or robbery or treason, it may also punish those who induce or counsel the commission of such crimes. Any other view makes a mockery of the state's power to declare and punish offences. And what the state may do to prevent the incitement of serious crimes which are universally condemned, it may also do to prevent the incitement of lesser crimes, or of those in regard to the bad tendency of which public opinion is divided. That is, if the state may punish John for burning straw in an alley, it may also constitutionally punish Frank for inciting John to do it, though Frank did so by speech or writing. And if, in 1857, the United States could punish John for helping a fugitive slave to escape, it could also punish Frank for inducing John to do this, even though a large section of public opinion might applaud John and condemn the Fugitive Slave Law.

Passage B

(from "Freedom of Speech in War Time" by Zechariah Chafee, Jr. written in 1919, published in Harvard Law Review Vol. 32 No. 8)

The true boundary line of the First Amendment can be fixed only when Congress and the courts realize that the principle on which speech is classified as lawful or unlawful involves the balancing against each other of two very important social interests, in public safety and in the search for truth. Every reasonable attempt should be made to maintain both interests unimpaired, and the great interest in free speech should be sacrificed only when the interest in public safety is really imperiled, and not, as most men believe, when it is barely conceivable that it may be slightly affected. In war time, therefore, speech should be unrestricted by the censorship or by punishment, unless it is clearly liable to cause direct and dangerous interference with the conduct of the war.

Thus our problem of locating the boundary line of free speech is solved. It is fixed close to the point where words will give rise to unlawful acts. We cannot define the right of free speech with the precision of the Rule against Perpetuities or the Rule in Shelley's Case, because it involves national policies which are much more flexible than private property, but we can establish a workable principle of classification in this method of balancing and this broad test of certain danger. There is a similar balancing in the determination of what is "due process of law." And we can with certitude declare that the First Amendment forbids the punishment of words merely for their injurious tendencies. The history of the Amendment and the political function of free speech corroborate each other and make this conclusion plain.

14. Which one of the following questions is central to both passages?
    a. What is the interpretation of the first amendment and its limitations?
    b. Do people want absolute liberty or do they only want liberty for a certain purpose?
    c. What is the true definition of freedom of speech in a democracy?
    d. How can we find an appropriate boundary of freedom of speech during wartime?

15. The authors of the two passages would be most likely to disagree over which of the following?
    a. A man is thrown in jail due to his provocation of violence in Washington D.C. during a riot.
    b. A man is thrown in jail for stealing bread for his starving family, and the judge has mercy for him and lets him go.
    c. A man is thrown in jail for encouraging a riot against the U.S. government for the wartime tactics although no violence ensues.
    d. A man is thrown in jail because he has been caught as a German spy working within the U.S. army.

16. The relationship between Passage *A* and Passage *B* is most analogous to the relationship between the documents described in which of the following?
    a. A journal article in the Netherlands about the law of euthanasia that cites evidence to support only the act of passive euthanasia as an appropriate way to die; a journal article in the Netherlands about the law of euthanasia that cites evidence to support voluntary euthanasia in any aspect.
    b. An article detailing the effects of radiation in Fukushima; a research report describing the deaths and birth defects as a result of the hazardous waste dumped on the Somali Coast.
    c. An article that suggests that labor laws during times of war should be left up to the states; an article that showcases labor laws during the past that have been altered due to the current crisis of war.
    d. A research report arguing that the leading cause of methane emissions in the world is from agriculture practices; an article citing that the leading cause of methane emissions in the world is from the transportation of coal, oil, and natural gas.

17. The author uses the examples in the last lines of Passage *A* in order to do what?
    a. To demonstrate different types of crimes for the purpose of comparing them to see by which one the principle of freedom of speech would become objectionable
    b. To demonstrate that anyone who incites a crime, despite the severity or magnitude of the crime, should be held accountable for that crime in some degree
    c. To prove that the definition of "freedom of speech" is altered depending on what kind of crime is being committed
    d. To show that some crimes are in the best interest of a nation and should not be punishable if they are proven to prevent harm to others

18. Which of the following, if true, would most seriously undermine the claim proposed by the author in Passage *A* that if the state can punish a crime then it can punish the incitement of that crime?

a. The idea that human beings are able and likely to change their mind between the utterance and execution of an event that may harm others
b. The idea that human beings will always choose what they think is right based on their cultural upbringing
c. The idea that the limitation of free speech by the government during wartime will protect the country from any group that causes a threat to that country's freedom
d. The idea that those who support freedom of speech probably have intentions of subverting the government

19. What is the primary purpose of the second passage?

a. To analyze the First Amendment in historical situations in order to make an analogy to the current war at hand in the nation
b. To demonstrate that the boundaries set during wartime are different from that when the country is at peace, and that we should change our laws accordingly
c. To offer the idea that during wartime, the principle of freedom of speech should be limited to that of even minor utterances in relation to a crime
d. To call upon the interpretation of freedom of speech to be already evident in the First Amendment and to offer a clear perimeter of the principle during war time

20. Which of the following words, if substituted for the word *malcontent* in Passage *A*, would LEAST change the meaning of the sentence?

a. Grievance
b. Cacophony
c. Anecdote
d. Residual

The next seven questions are from <u>Rhetoric and Poetry in the Renaissance: A Study of Rhetorical Terms in English Renaissance Literary Criticism</u> by DL Clark

> To the Greeks and Romans rhetoric meant the theory of oratory. As a pedagogical mechanism it endeavored to teach students to persuade an audience. The content of rhetoric included all that the ancients had learned to be of value in persuasive public speech. It taught how to work up a case by drawing valid inferences from sound evidence, how to organize this material in the most persuasive order, how to compose in clear and harmonious sentences. Thus to the Greeks and Romans rhetoric was defined by its function of discovering means to persuasion and was taught in the schools as something that every free-born man could and should learn.
>
> In both these respects the ancients felt that poetics, the theory of poetry, was different from rhetoric. As the critical theorists believed that the poets were inspired, they endeavored less to teach men to be poets than to point out the excellences which the poets had attained. Although these critics generally, with the exceptions of Aristotle and Eratosthenes, believed the greatest value of poetry to be in the teaching of morality, no one of them endeavored to define poetry, as they did rhetoric, by its purpose. To Aristotle, and centuries later to Plutarch, the distinguishing mark of poetry was imitation. Not until the renaissance did critics define poetry as an art of imitation endeavoring to inculcate morality . . .
>
> The same essential difference between classical rhetoric and poetics appears in the content of classical poetics. Whereas classical rhetoric deals with speeches which might be delivered to

convict or acquit a defendant in the law court, or to secure a certain action by the deliberative assembly, or to adorn an occasion, classical poetic deals with lyric, epic, and drama. It is a commonplace that classical literary critics paid little attention to the lyric. It is less frequently realized that they devoted almost as little space to discussion of metrics. By far the greater bulk of classical treatises on poetics is devoted to characterization and to the technique of plot construction, involving as it does narrative and dramatic unity and movement as distinct from logical unity and movement.

21. What does the author say about one way in which the purpose of poetry changed for later philosophers?

a. The author says that at first, poetry was not defined by its purpose but was valued for its ability to be used to teach morality. Later, some philosophers would define poetry by its ability to instill morality. Finally, during the renaissance, poetry was believed to be an imitative art, but was not necessarily believed to instill morality in its readers.
b. The author says that the classical understanding of poetry dealt with its ability to be used to teach morality. Later, philosophers would define poetry by its ability to imitate life. Finally, during the renaissance, poetry was believed to be an imitative art that instilled morality in its readers.
c. The author says that at first, poetry was thought to be an imitation of reality, then later philosophers valued poetry more for its ability to instill morality.
d. The author says that the classical understanding of poetry was that it dealt with the search for truth through its content; later, the purpose of poetry would be through its entertainment.

22. What does the author of the passage say about classical literary critics in relation to poetics?

a. That rhetoric was more valued than poetry because rhetoric had a definitive purpose to persuade an audience, and poetry's wavering purpose made it harder for critics to teach.
b. That although most poetry was written as lyric, epic, or drama, the critics were most focused on the techniques of lyric and epic and their performance of musicality and structure.
c. That although most poetry was written as lyric, epic, or drama, the critics were most focused on the techniques of the epic and drama and their performance of structure and character.
d. That the study of poetics was more pleasurable than the study of rhetoric due to its ability to assuage its audience, and the critics therefore focused on what poets did to create that effect.

23. What is the primary purpose of this passage?

a. To contemplate the differences between classical rhetoric and poetry and to consider their purposes in a particular culture.
b. To inform the readers of the changes in poetic critical theory throughout the years and to contrast those changes to the solidity of rhetoric.
c. To educate the audience on rhetoric by explaining the historical implications of using rhetoric in the education system.
d. To convince the audience that poetics is a subset of rhetoric as viewed by the Greek and Roman culture.

24. The word *inculcate* in paragraph two can be best interpreted as referring to which one of the following?

a. Imbibe
b. Instill
c. Implode
d. Inquire

25. Which of the following most closely resembles the way in which the passage is structured?

a. The first paragraph presents us with an issue. The second paragraph offers a solution to the problem. The third paragraph summarizes the first two paragraphs.

b. The first paragraph presents us with definitions and examples of a particular subject. The second paragraph presents a second subject in the same way. The third paragraph offers a contrast of the two subjects.

c. The first paragraph presents an inquiry. The second paragraph explains the details of that inquiry. The last paragraph offers a solution.

d. The first paragraph presents us with two subjects alongside definitions and examples. The second paragraph presents us with a comparison of the two subjects. The third paragraph presents us with a contrast of the two subjects.

26. Given the author's description of the content of rhetoric in the first paragraph, which one of the following is most analogous to what it taught? (The sentence is shown below.)

It taught how to work up a case by drawing valid inferences from sound evidence, how to organize this material in the most persuasive order, how to compose in clear and harmonious sentences.

a. As a musician, they taught me that the end product of the music is everything—what I did to get there was irrelevant, whether it was my ability to read music or the reliance on my intuition to compose.

b. As a detective, they taught me that time meant everything when dealing with a new case, that the simplest explanation is usually the right one, and that documentation is extremely important to credibility.

c. As a writer, they taught me the most important thing about writing was consistently showing up to the page every single day, no matter where my muse was.

d. As a football player, they taught me how to understand the logistics of the game, how my placement on the field affected the rest of the team, and how to run and throw with a mixture of finesse and strength.

27. Which of the following words, if substituted for the word *treatises* in paragraph three, would LEAST change the meaning of the sentence?

a. Commentary

b. Encyclopedias

c. Sermons

d. Anthems

The next four questions are based on the following passage. It is from Oregon, Washington, and Alaska. Sights and Scenes for the Tourist, written by E.L. Lomax in 1890:

Portland is a very beautiful city of 60,000 inhabitants and situated on the Willamette river twelve miles from its junction with the Columbia. It is perhaps true of many of the growing cities of the West, that they do not offer the same social advantages as the older cities of the East. But this is principally the case as to what may be called boom cities, where the larger part of the population is of that floating class which follows in the line of temporary growth for the purposes of speculation, and in no sense applies to those centers of trade whose prosperity is based on the solid foundation of legitimate business. As the metropolis of a vast section of country, having broad agricultural valleys filled with improved farms, surrounded by mountains rich in mineral wealth, and boundless forests of as fine timber as the world produces, the cause of Portland's growth and prosperity is the trade which it has as the center of collection and distribution of this great wealth of natural resources, and it has attracted, not the boomer and speculator, who find their profits in the wild excitement of the boom, but the merchant, manufacturer, and investor, who seek the surer if slower channels of legitimate business and investment. These have come from the East, most of them within the last few years. They came as seeking a better and wider field to engage in the same occupations they had followed in their Eastern homes, and bringing with them all the love of polite life which they had acquired there, have established here a new society, equaling in all respects that which they left behind. Here are as fine churches, as complete a system of schools, as fine residences, as great a love of music and art, as can be found at any city of the East of equal size.

But while Portland may justly claim to be the peer of any city of its size in the United States in all that pertains to social life, in the attractions of beauty of location and surroundings it stands without its peer. The work of art is but the copy of nature. What the residents of other cities see but in the copy, or must travel half the world over to see in the original, the resident of Portland has at its very door.

The city is situate on a gently-sloping ground, with, on the one side, the river, and on the other a range of hills, which, within easy walking distance, rise to an elevation of a thousand feet above the river, affording a most picturesque building site. From the very streets of the thickly settled portion of the city, the Cascade Mountains, with the snow-capped peaks of Hood, Adams, St. Helens, and Rainier, are in plain view.

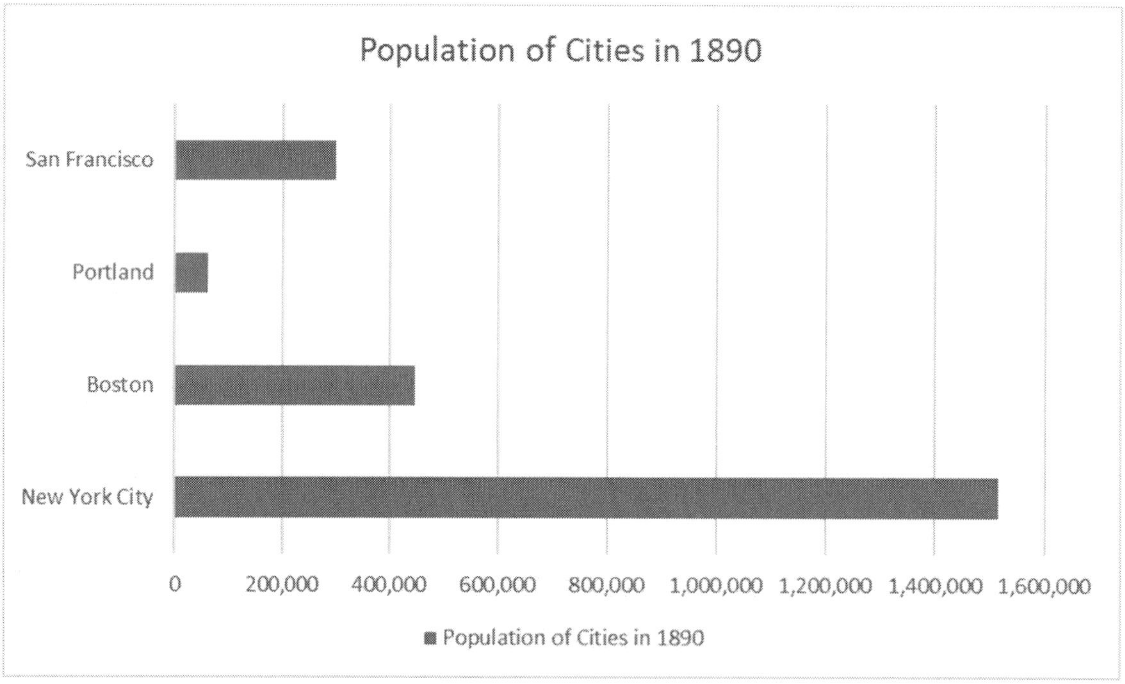

28. What is a characteristic of a "boom city," as indicated by the passage?
   a. A city that is built on solid business foundation of mineral wealth and farming.
   b. An area of land on the west coast that quickly becomes populated by residents from the east coast.
   c. A city that, due to the hot weather and dry climate, catches fire frequently, resulting in a devastating population drop.
   d. A city whose population is made up of people who seek quick fortunes rather than building a solid business foundation.

29. By stating that "they do not offer the same social advantages as the older cities of the East" in the first paragraph, the author most likely intends to suggest that:
   a. Inhabitants who reside in older cities in the East are much more social than inhabitants who reside in newer cities in the West because of background and experience.
   b. Cities in the West have no culture compared to the East because the culture in the East comes from European influence.
   c. Cities in the East are older than cities in the West, and older cities always have better culture than newer cities.
   d. Since cities in the West are newly established, it takes them a longer time to develop cultural roots and societal functions than those cities that are already established in the East.

30. Based on the information at the end of paragraph 1, what would the author say of Portland?
   a. It has twice as much culture as the cities in the East.
   b. It has as much culture as the cities in the East.
   c. It doesn't have as much culture as cities in the East.
   d. It doesn't have as much culture as cities in the West.

31. How many more citizens did San Francisco have than Portland in 1890?
   a. Approximately 240,000
   b. Approximately 500,000
   c. Approximately 1,000,000
   d. Approximately 1,500,000

The following four questions are based on the excerpt from Variation of Animals and Plants by Charles Darwin.

> Peach (Amygdalus persica).—In the last chapter I gave two cases of a peach-almond and a double-flowered almond which suddenly produced fruit closely resembling true peaches. I have also given many cases of peach-trees producing buds, which, when developed into branches, have yielded nectarines. We have seen that no less than six named and several unnamed varieties of the peach have thus produced several varieties of nectarine. I have shown that it is highly improbable that all these peach-trees, some of which are old varieties, and have been propagated by the million, are hybrids from the peach and nectarine, and that it is opposed to all analogy to attribute the occasional production of nectarines on peach-trees to the direct action of pollen from some neighbouring nectarine-tree. Several of the cases are highly remarkable, because, firstly, the fruit thus produced has sometimes been in part a nectarine and in part a peach; secondly, because nectarines thus suddenly produced have reproduced themselves by seed; and thirdly, because nectarines are produced from peach-trees from seed as well as from buds. The seed of the nectarine, on the other hand, occasionally produces peaches; and we have seen in one instance that a nectarine-tree yielded peaches by bud-variation. As the peach is certainly the oldest or primary variety, the production of peaches from nectarines, either by seeds or buds, may

perhaps be considered as a case of reversion. Certain trees have also been described as indifferently bearing peaches or nectarines, and this may be considered as bud-variation carried to an extreme degree.

The grosse mignonne peach at Montreuil produced "from a sporting branch" the grosse mignonne tardive, "a most excellent variety," which ripens its fruit a fortnight later than the parent tree, and is equally good. This same peach has likewise produced by bud-variation the early grosse mignonne. Hunt's large tawny nectarine "originated from Hunt's small tawny nectarine, but not through seminal reproduction."

32. Which statement is not a detail from the passage?
   a. At least six named varieties of the peach have produced several varieties of nectarine.
   b. It is not probable that all of the peach-trees mentioned are hybrids from the peach and nectarine.
   c. An unremarkable case is the fact that nectarines are produced from peach-trees from seed as well as from buds.
   d. The production of peaches from nectarines might be considered a case of reversion.

33. What is the meaning of the word "propagated" in the first paragraph of this passage?
   a. Multiplied
   b. Diminished
   c. Watered
   d. Uprooted

34. Which of the following most closely reveals the author's tone in this passage?
   a. Enthusiastic
   b. Objective
   c. Critical
   d. Desperate

35. Which of the following is an accurate paraphrasing of the following phrase?

> Certain trees have also been described as indifferently bearing peaches or nectarines, and this may be considered as bud-variation carried to an extreme degree.

   a. Some trees are described as bearing peaches and some trees have been described as bearing nectarines, but individually the buds are extreme examples of variation.
   b. One way in which bud-variation is said to be carried to an extreme degree is when specific trees have been shown to casually produce peaches or nectarines.
   c. Certain trees are indifferent to bud-variation, as recently shown in the trees that produce both peaches and nectarines in the same season.
   d. Nectarines and peaches are known to have cross-variation in their buds, which indifferently bears other sorts of fruit to an extreme degree.

The following four questions are based on the excerpt from A Christmas Carol by Charles Dickens.

Meanwhile the fog and darkness thickened so, that people ran about with flaring links, proffering their services to go before horses in carriages, and conduct them on their way. The ancient tower of a church, whose gruff old bell was always peeping slily down at Scrooge out of a Gothic window in the wall, became invisible, and struck the hours and quarters in the clouds, with tremulous vibrations afterwards as if its teeth were chattering in its frozen head up there. The cold became intense. In the main street, at the corner of the court, some labourers were repairing the gas-pipes, and had lighted a great fire in a brazier, round which a party of ragged men and boys were gathered: warming their hands and winking their eyes before the blaze in rapture. The water-plug being left in solitude, its overflowings sullenly congealed, and turned to misanthropic ice. The brightness of the shops where holly sprigs and berries crackled in the lamp heat of the windows, made pale faces ruddy as they passed. Poulterers' and grocers' trades became a splendid joke; a glorious pageant, with which it was next to impossible to believe that such dull principles as bargain and sale had anything to do. The Lord Mayor, in the stronghold of the mighty Mansion House, gave orders to his fifty cooks and butlers to keep Christmas as a Lord Mayor's household should; and even the little tailor, whom he had fined five shillings on the previous Monday for being drunk and bloodthirsty in the streets, stirred up to-morrow's pudding in his garret, while his lean wife and the baby sallied out to buy the beef.

Foggier yet, and colder. Piercing, searching, biting cold. If the good Saint Dunstan had but nipped the Evil Spirit's nose with a touch of such weather as that, instead of using his familiar weapons, then indeed he would have roared to lusty purpose. The owner of one scant young nose, gnawed and mumbled by the hungry cold as bones are gnawed by dogs, stopped down at Scrooge's keyhole to regale him with a Christmas carol: but at the first sound of

"God bless you, merry gentleman! May nothing you dismay"

Scrooge seized the ruler with such energy of action, that the singer fled in terror, leaving the keyhole to the fog and even more congenial frost.

36. In the context in which it appears, *congealed* most nearly means which of the following?
   a. Burst
   b. Loosened
   c. Shrank
   d. Thickened

37. Which of the following can NOT be inferred from the passage?
   a. The season of this narrative is in the winter time.
   b. The majority of the narrative is located in a bustling city street.
   c. This passage takes place during the night time.
   d. The Lord Mayor is a wealthy person within the narrative.

38. According to the passage, the poulterers and grocers
   a. were so poor in the quality of their products that customers saw them as a joke.
   b. put on a pageant in the streets every year for Christmas to entice their customers.
   c. did not believe in Christmas so they refused to participate in the town parade.
   d. set their shops up to be entertaining public spectacles rather than a dull trade exchange.

39. The author's depiction of the scene in the last few paragraphs does all of the following EXCEPT:
   a. Offer an allusion to religious affiliation in England.
   b. Attempt to evoke empathy for the character of Scrooge.
   c. Provide a palpable experience through the use of imagery and diction.
   d. Depict Scrooge as an uncaring, terrifying character to his fellows.

The following four questions are based on the book <u>On the Trail</u> by Lina Beard and Adelia Belle Beard.

For any journey, by rail or by boat, one has a general idea of the direction to be taken, the character of the land or water to be crossed, and of what one will find at the end. So it should be in striking the trail. Learn all you can about the path you are to follow. Whether it is plain or obscure, wet or dry; where it leads; and its length, measured more by time than by actual miles. A smooth, even trail of five miles will not consume the time and strength that must be expended upon a trail of half that length which leads over uneven ground, varied by bogs and obstructed by rocks and fallen trees, or a trail that is all up-hill climbing. If you are a novice and accustomed to walking only over smooth and level ground, you must allow more time for covering the distance than an experienced person would require and must count upon the expenditure of more strength, because your feet are not trained to the wilderness paths with their pitfalls and traps for the unwary, and every nerve and muscle will be strained to secure a safe foothold amid the tangled roots, on the slippery, moss-covered logs, over precipitous rocks that lie in your path. It will take time to pick your way over boggy places where the water oozes up through the thin, loamy soil as through a sponge; and experience alone will teach you which hummock of grass or moss will make a safe stepping-place and will not sink beneath your weight and soak your feet with hidden water. Do not scorn to learn all you can about the trail you are to take . . . It is not that you hesitate to encounter difficulties, but that you may prepare for them. In unknown regions take a responsible guide with you, unless the trail is short, easily followed, and a frequented one. Do not go alone through lonely places; and, being on the trail, keep it and try no explorations of your own, at least not until you are quite familiar with the country and the ways of the wild.

Blazing the Trail

A woodsman usually blazes his trail by chipping with his axe the trees he passes, leaving white scars on their trunks, and to follow such a trail you stand at your first tree until you see the blaze on the next, then go that and look for the one farther on; going in this way from tree to tree you keep the trail though it may, underfoot, be overgrown and indistinguishable.

If you must make a trail of your own, blaze it as you go by bending down and breaking branches of trees, underbrush, and bushes. Let the broken branches be on the side of bush or tree in the direction you are going, but bent down away from that side, or toward the bush, so that the lighter underside of the leaves will show and make a plain trail. Make these signs conspicuous and close together, for in returning, a dozen feet without the broken branch will sometimes confuse you, especially as everything has a different look when seen from the opposite side. By this same token it is a wise precaution to look back frequently as you go and impress the homeward-bound landmarks on your memory. If in your wanderings you have branched off and made ineffectual or blind trails which lead nowhere, and, in returning to camp, you are led astray by one of them, do not leave the false trail and strike out to make a new one, but turn back and follow the false trail to its beginning, for it must lead to the true trail again. Don't lose sight of your broken branches.

40. In the image, what part of the text is the girl most likely emulating?
    a. Building a trap
    b. Setting up camp
    c. Blazing the trail
    d. Picking berries to eat

41. According to the passage, what does the author say about unknown regions?
    a. You should try and explore unknown regions in order to learn the land better.
    b. Unless the trail is short or frequented, you should take a responsible guide with you.
    c. All unknown regions will contain pitfalls, traps, and boggy places.
    d. It's better to travel unknown regions by rail rather than by foot.

42. Which statement is NOT a detail from the passage?
    a. Learning about the trail beforehand is imperative.
    b. Time will differ depending on the land.
    c. Once you are familiar with the outdoors you can go places on your own.
    d. Be careful for wild animals on the trail you are on.

43. In the last paragraph, which of the following does the author suggest when being led astray by a false trail?

a. Bend down and break the branches off trees, underbrush, and bushes.
b. Ignore the false trail and strike out to make a new one.
c. Follow the false trail back to its beginning so that you can rediscover the real trail.
d. Make the signs conspicuous so that you won't be confused when you turn around.

The following is from Poems by Alexander Pushkin *by Ivan Panin. The next four questions are based on the following passage:*

> I do not believe there are as many as five examples of deviation from the literalness of the text. Once only, I believe, have I transposed two lines for convenience of translation; the other deviations are (*if* they are such) a substitution of an *and* for a comma in order to make now and then the reading of a line musical. With these exceptions, I have sacrificed *everything* to faithfulness of rendering. My object was to make Pushkin himself, without a prompter, speak to English readers. To make him thus speak in a foreign tongue was indeed to place him at a disadvantage; and music and rhythm and harmony are indeed fine things, but truth is finer still. I wished to present not what Pushkin would have said, or should have said, if he had written in English, but what he does say in Russian. That, stripped from all ornament of his wonderful melody and grace of form, as he is in a translation, he still, even in the hard English tongue, soothes and stirs, is in itself a sign that through the individual soul of Pushkin sings that universal soul whose strains appeal forever to man, in whatever clime, under whatever sky.
>
> I ask, therefore, no forgiveness, no indulgence even, from the reader for the crudeness and even harshness of the translation, which, I dare say, will be found in abundance by those who *look* for something to blame. Nothing of the kind is necessary. I have done the only thing there was to be done. Nothing more *could* be done (I mean by me, of course), and if critics still demand more, they must settle it not with me, but with the Lord Almighty, who in his grim, yet arch way, long before critics appeared on the stage, hath ordained that it shall be impossible for a thing to be and not to be at the same time.
>
> I have therefore tried neither for measure nor for rhyme. What I have done was this: I first translated each line word for word, and then by reading it aloud let mine ear arrange for me the words in such a way as to make some kind of rhythm. Where this could be done, I was indeed glad; where this could not be done, I was not sorry. It is idle to regret the impossible.

44. From clues in this passage, what type of work is the author doing?

a. Translation work
b. Criticism
c. Historical validity
d. Writing a biography

45. Where would you most likely find this passage in a text?

a. Appendix
b. Table of contents
c. First chapter
d. Preface

46. According to the author, what is the most important aim of translation work?
    a. To retain the beauty of the work.
    b. To retain the truth of the work.
    c. To retain the melody of the work.
    d. To retain the form of the work.

47. What is the author trying to express in the second paragraph?
    a. The author is trying to say that he is not regretful for any crudeness on the part of the text because the translation was done to the best of the author's ability.
    b. The author is asking for forgiveness because his translation may be crude in some places, and it is not true to the meter or rhyme of the original text.
    c. The author wants to express that critics are cast out in the presence of God because their work is insincere and pointless.
    d. The author is saying that he is not afraid to fight somebody if they criticize his work or the work of his peers.

The following is entitled Architecture and Democracy by Claude Bragdon. The next six questions are based on the following passage.

The world war represents not the triumph, but the birth of democracy. The true ideal of democracy—the rule of a people by the *demos*, or group soul—is a thing unrealized. How then is it possible to consider or discuss an architecture of democracy—the shadow of a shade? It is not possible to do so with any degree of finality, but by an intention of consciousness upon this juxtaposition of ideas—architecture and democracy—signs of the times may yield new meanings, relations may emerge between things apparently unrelated, and the future, always existent in every present moment, may be evoked by that strange magic which resides in the human mind.

Architecture, at its worst as at its best, reflects always a true image of the thing that produced it; a building is revealing even though it is false, just as the face of a liar tells the thing his words endeavor to conceal. This being so, let us make such architecture as is ours declare to us our true estate.

The architecture of the United States, from the period of the Civil War, up to the beginning of the present crisis, everywhere reflects a struggle to be free of a vicious and depraved form of feudalism, grown strong under the very ægis of democracy. The qualities that made feudalism endeared and enduring; qualities written in beauty on the cathedral cities of mediaeval Europe—faith, worship, loyalty, magnanimity—were either vanished or banished from this pseudo-democratic, aridly scientific feudalism, leaving an inheritance of strife and tyranny—a strife grown mean, a tyranny grown prudent, but full of sinister power the weight of which we have by no means ceased to feel.

Power, strangely mingled with timidity; ingenuity, frequently misdirected; ugliness, the result of a false ideal of beauty—these in general characterize the architecture of our immediate past; an architecture "without ancestry or hope of posterity," an architecture devoid of coherence or conviction; willing to lie, willing to steal. What impression such a city as Chicago or Pittsburgh might have made upon some denizen of those cathedral-crowned feudal cities of the past we do not know. He would certainly have been amazed at its giant energy, and probably revolted at its grimy dreariness. We are wont to pity the mediaeval man for the dirt he lived in, even while smoke greys our sky and dirt permeates the very air we breathe: we think of castles as grim and cathedrals as dim, but they were beautiful and gay with color compared with the grim, dim canyons of our city streets.

48. Which of the following does the author NOT consider to be a characteristic of modern architecture?
   a. Power, strangely mingled with timidity
   b. Ugliness as a result of the false ideal of beauty
   c. Giant energy with grimy and dreariness
   d. Cathedral-crowned grim and dim castles

49. By stating that "Architecture, at its worst as at its best, reflects always a true image of the thing that produced it," the author most likely intends to suggest that:
   a. People always create buildings to look like themselves.
   b. Architecture gets more grim, drab, and depressing as the years go by.
   c. Architecture reflects—in shape, color, and form—the attitude of the society which built it.
   d. Modern architecture is a lot like democracy because it is uniform yet made up of more pieces than your traditional architecture.

50. The author refers to "mediaeval man" in the fourth paragraph in order to:
   a. Make the audience look at feudalism with a sense of nostalgia and desire.
   b. Make the audience feel gratitude for modern comforts such as architecture.
   c. Make the audience realize the irony produced from pitying him.
   d. Make the audience look back at feudalism as a time which was dark and dreary.

51. Based on the discussion in paragraph 3, which of the following would be considered architecture from medieval Europe?
   a. Canyon
   b. Castle
   c. Factory
   d. Skyscraper

52. The author's attitude toward modern architecture can best be characterized as:
   a. Narcissistic
   b. Aggrieved
   c. Virtuous
   d. Sarcastic

Writing and Language

Questions 1–5 are based on the following passage.

> Nobody knew where she went. She wasn't in the living room, the dining room, or the kitchen. She wasn't underneath the staircase, and she certainly wasn't in her bedroom. The bathroom was thoroughly checked, and the closets were emptied. Where could she have possibly gone? Outside was dark, cold, and damp. Each person took a turn calling for her. "Eva! Eva!" (1) they'd each cry out, but not a sound was heard in return. The littlest sister in the house began to cry. "Where could my cat have gone?" Everyone just looked at each other (2) in utter silence. Lulu was only five years old and brought this kitten home only six weeks ago and she was so fond of her. (3) Every morning, Lulu would comb her fur, feed her, and play a cat-and-mouse game with her before she left for school. At night, Eva would curl up on the pillow next to Lulu and stay with her across the night (4). They were already becoming best friends. But this evening, right after dinner, Eva snuck outside when Lulu's brother came in from the yard. It was already dark out and Eva ran so quickly that nobody could see where she went.

It was really starting to feel hopeless when all of a sudden Lulu had an idea. She remembered how much Eva loved the cat-and-mouse game, so she decided to go outside and pretend that she was playing with Eva. Out she went. She curled up on the grass and began to play cat and mouse. Within seconds, Lulu could see shining eyes in the distance. The eyes got closer, and closer, and closer. Lulu continued to play <u>in order that</u> (5) she wouldn't scare the kitten away. In less than five minutes, her beautiful kitten had crawled right up to her lap as if asking to join in the game. Eva was back, and Lulu couldn't have been more thrilled! Gently, she scooped up her kitten, grabbed the cat-and-mouse game, and headed back inside, grinning from ear to ear. That night, Eva and Lulu snuggled right back down again, side by side, and had a very restful sleep.

1. Choose the best replacement punctuation.
   a. NO CHANGE
   b. 'Eva! Eva!'
   c. 'Eva, Eva'!
   d. "Eva, Eva"!

2. Choose the best replacement word or phrase.
   a. NO CHANGE
   b. one another
   c. themselves
   d. each one

3. Choose the best replacement sentence.
   a. NO CHANGE
   b. Lulu brought this kitten home only six weeks ago and she was so fond of her
   c. Lulu brought this kitten home only six weeks ago, and she was so fond of her
   d. Lulu was only five years old, and brought this kitten home only six weeks ago, and she was so fond of her

4. Choose the best replacement word or phrase.
   a. NO CHANGE
   b. below the night
   c. through the night
   d. amidst the night

5. Choose the best replacement word or phrase.
   a. NO CHANGE
   b. so,
   c. so that
   d. so that,

Questions 6–10 are based on the following passage.

In our busy family, there are chores that each member must do every day. There are chores for our daughter, for me, and for my husband. There is always so much going on in our family and so many scheduled events every week that without a set list of <u>chores and</u> (6) without ensuring the chores are completed, our household would be utterly chaotic.

I work days, from home. My husband works evenings about a half-hour away, and our <u>13-year-old</u> (7) daughter, Niamh, attends a homeschooling program. She is an avid hockey player and the third-highest scorer on the team. She simply lives for hockey. Three times a week, she has hockey practice, and once, approximately, a week (8), she has a hockey game. Niamh is responsible for washing her jerseys and her hockey pants and making sure all her gear is properly stored after every practice and game in the garage. Occasionally she forgets and leaves it all in the middle of the dining room floor. Niamh must also clean the kitchen every evening after dinner, empty the dishwasher, and feed her two guinea pigs and her cat. Every Saturday, we also give Niamh an allowance for doing extra chores. She sweeps and washes our floors, cleans windows, vacuums, and tidies up the bathrooms.

As for me, I am responsible for walking and feeding the dogs, preparing the lunches and dinners, and keeping up with the laundry. My husband takes out the trash and recycling, keeps up with the yard work, and pays all the bills. Both my husband and myself (9) drive our daughter to friends' houses, school, the arena, the grocery store, medical appointments, and more.

We very rarely seem to get a break, but we are always happy and healthy, and that's what really matters. So, now that you know the breakdown of chores between our daughter, my husband, and me, who do you think has more (10) chores?

6. Choose the best replacement.
   a. NO CHANGE
   b. chores, and,
   c. chores and,
   d. chores, and

7. Choose the best replacement.
   a. NO CHANGE
   b. 13 year old
   c. 13-year old
   d. 13 year-old

8. Choose the best replacement.
   a. NO CHANGE
   b. approximately once a week
   c. approximately bi-weekly
   d. once a week

9. Choose the best replacement.
   a. NO CHANGE
   b. I
   c. me
   d. myself,

10. Choose the best replacement.
   a. NO CHANGE
   b. the most
   c. lesser
   d. less

Questions 11–15 are based on the following passage.

Imagine yourself in a secluded room. There is no one to talk to. The room has a bed, a desk, and a computer, and its one window faces the yard. Every day, you must work in this isolated room, away from everyone. There isn't a sound heard accept (11) for the birds outside (12) and your own hands tapping on the computer keyboard. Imagine having to work in that somber environment every day. The only breaks you receive are to go to the bathroom or grab a quick bite to eat. It feels like solitary confinement, and you start to feel yourself slipping ever-so (13) slowly into a depressed state. What should you do? If you don't get the work done, you won't get paid.

Suddenly, you have an idea. You decide that at least two days a week, you can pack up and work at the library. Although it is quiet at the library, there are people coming and going. You will be able to hear the librarians helping students and visitors, and you can people-watch. Feeling optimistic, the day suddenly makes you smile. (14) When the morning arrives, you pack up your computer, notebooks, calendar, phone, and you head out the door. You arrive at the library just as the doors are opening. You find a quiet place to set up and (15) you settle down into your work day. At first, you are thrilled with the new scenery and you love the extra space you have to work with. But before long, you realize your work isn't getting done. You spent so much time watching the different people coming and going and daydreaming about what it would be like to casually read a book that you stopped working altogether! Perhaps the library wasn't the ideal spot for your after all.

You pack up your computer, your notebooks, your calendar, and your phone, and you head back out the door, back to the solitary room with the bed, the desk, the computer, and the window, and you settle down once again, into your work day.

11. Choose the best replacement.
   a. NO CHANGE
   b. exceptionally
   c. except
   d. expect

12. Choose the best replacement.
   a. NO CHANGE
   b. birds chirping outside
   c. birds, outside
   d. birds, chirping outside

13. Choose the best replacement.
   a. NO CHANGE
   b. ever-so,
   c. ever so
   d. everso

14. Choose the best replacement word or phrase.
   a. NO CHANGE
   b. the day no longer seems so sad.
   c. the day brightens your mood.
   d. you smile for the rest of the day.

15. Choose the best replacement.
   a. NO CHANGE
   b. set up, and
   c. set up; and
   d. set up. And

Questions 16–20 are based on the following passage.

> Why must I always be last? I'm the last one in line at the cafeteria, I'm the last one to my locker, and I'm always last (16) one to class. For once, I'd really like to be first. I would even be happy being right smack dab in the middle!
>
> Having also been the last of all the siblings, my older sisters always teased me. (17) I was the last to be potty-trained, the last to get a big girl's bed, and the last to start school. When we entered the teenage years, I was the last to get my driver's license and the last to dye my hair—I went with purple. My older sisters did spoil me, though. When they got their driver's licenses, they would take me out for ice cream or to the movies, and I even got to tag along from time to time when they went out on dates. I guess they thought I was cute or something. Each time I would complain about being last, my mother would say, "Don't ever worry about being last. That means you're the best!" I never quite knew what that meant. My father's famous line was, 'Last, but not least'! (18) and I'm still trying to figure *that* one out too.
>
> Of course, I'm not being completely honest with you and (19) I've never been the last in my class. In fact, I always seem to score in the top tenth percentile which isn't too bad at all. I have come in first, second, and even third in different swimming events, and when it comes to dinner, I am usually the *first* to finish my plate. And now that I think about it, maybe being last doesn't really matter anyway. After all (20), being the youngest in my family means that I get away with a lot too. I don't have to do a lot of the chores my older sisters have to do. Yes, come to think of it, maybe being last isn't so bad after all.

16. Choose the best replacement.
   a. NO CHANGE
   b. the last
   c. late
   d. the last,

17. Choose the best replacement.
   a. NO CHANGE
   b. my older sisters would always tease me.
   c. I was always teased by my older sisters.
   d. my older sisters, always, teased me.

18. Choose the best replacement.
   a. NO CHANGE
   b. 'Last but not least!'
   c. "Last, but not least"!
   d. "Last but not least!"

19. Choose the best replacement.
   a. NO CHANGE
   b. ;
   c. even though
   d. but

20. Choose the best replacement.
   a. NO CHANGE
   b. Therefore,
   c. However,
   d. On the contrary,

Questions 21–25 are based on the following passage.

> Café Adriatico was the best place in town to go for authentic Italian cuisine. The entire restaurant consisted of the kitchen in the far back of the restaurant, one small bathroom, and a quaint dining area that comfortably held ten tables. There was always Italian classical music playing in the background, and the atmosphere always put me in mind of being in my own grandparents (21) home.
>
> Angela, the sole owner and head chef, was a unique character. She loved her guests and treated them like her extended family. She would saunter (22) around from table to table, talking with the diners and occasionally pulling up a chair to gab a little longer. One famous story she used to tell was of her younger years in Italy, and how once she won a national beauty contest. (23) She wouldn't tell the story in an arrogant way. She would just matter-of-factly tell us what it was like when she was honored with this award. She was reminiscing about the good old days, I guess, and who doesn't like to do that?
>
> People didn't expect to be fed right away at Café Adriatico. Everything was cooked from scratch and made from the freshest and highest-quality of ingredients. Going to Café Adriatico was the equivalent of going out for the evening. Boy, was Angela something else! It wasn't the least bit uncommon to see her walking around the restaurant with a cigarette hanging out of her mouth and her tiny poodle walking along at her ankles. This doesn't (24) seem to bother any of the locals though. Angela was too well-liked (25). People went there for the comradery, the excellent cuisine, and the reminder of what it was like to be served an authentic Italian meal in the comfort of one's home. If you ever visit Hamilton, Ontario, Canada, be sure to visit Café Adriatico. And if Angela is still sauntering from table to table, please tell her I say hello!

21. Choose the best replacement.
   a. NO CHANGE
   b. grandparent's
   c. grandparents'
   d. grandparents's

22. Choose the best replacement.
   a. NO CHANGE
   b. sauntered
   c. was sauntering
   d. had sauntered

23. Choose the best replacement.
   a. NO CHANGE
   b. National Beauty Contest
   c. National beauty contest
   d. national Beauty contest

24. Choose the best replacement.
   a. NO CHANGE
   b. hadn't
   c. didn't
   d. never

25. Choose the best replacement.
   a. NO CHANGE
   b. well liked
   c. much liked
   d. liked

Questions 26–30 are based on the following passage.

> The school bell was about to ring at Juniper High, but Matt wasn't ready. He wasn't ready because he hadn't studied for today's science test in Mr. Jones (26) class. He planned on studying. He brought his textbook, notebook, and study guide home, but something unexpected happened last night that changed his plans. About an hour after he arrived home from school, Matt's dad walked in the door with a brand-new synthesizer, and that was the end of studying before it even began.
>
> Ever since Matt was five years old, he was obsessed with music—any kind of music. He would listen to his parents' kitchen radio for hours, play (27) their records and singing away to all the classics. When he was a teenager, he started collecting CDs, and within no time at all he had an impressive collection of artists. Therefore (28) what he really wanted was his very own synthesizer so he could create his own music—and that day had finally come! Matt forgot all about his science test and had begun (29) to experiment with his new toy immediately. He played for hours and tested every button on his synthesizer. Before he knew it, the clock struck midnight, and he fell into a deep sleep with music playing in his head.
>
> The bell rang. Matt slowly walked into school, wondering how he was going to pass this test. It was the last test of the semester, and it was the one that was supposed to bring up his average. If he failed this test, he failed (30) the semester. He went to his locker, grabbed what he needed, and headed off to class. Within minutes, Matt knew this was his lucky day. The classroom door opened, and a substitute teacher walked in—the test would be delayed by a day! Now all Matt had to do was go home and study, even though that synthesizer would be tempting him all night.

26. Choose the best replacement.
   a. NO CHANGE
   b. Jones's
   c. Jones'
   d. Jone's

27. Choose the best replacement.
    a. NO CHANGE
    b. playing
    c. played
    d. would be playing

28. Choose the best replacement.
    a. NO CHANGE
    b. However
    c. However,
    d. Nonetheless,

29. Choose the best replacement word or phrase.
    a. NO CHANGE
    b. was beginning
    c. begins
    d. began

30. Choose the best replacement phrase.
    a. NO CHANGE
    b. had failed
    c. was going to fail
    d. would fail

Questions 31–35 are based on the following passage.

Everybody has their own version of the perfect peanutbutter (31) and banana sandwich. Some will tell the type of bread you must use and whether the bread should be toasted. Some will tell you that you must cut off the crust, while others will tell you to keep it intact. Crunchy or smooth is always a heated debate, along with just how much peanut butter a sandwich needs.

I'm no different; I consider myself an expert when it comes to peanut butter and banana sandwiches. My grandmother started making me and my brothers peanut butter and banana sandwiches (32) before we could speak. I'd watch my grandmother in the kitchen closely. She would start by setting out the slices of bread along the counter, and then carefully add (33) one thin layer of butter to each slice. She would then start back at the top and add one fine layer of smooth peanut butter. Next was the final and best step. (34) Instead of slicing the bananas, my grandmother would mash them! She would take a bowl, place the bananas inside, and mash them up with the back of a fork! Then she would spread the slices of bread with the mashed banana. In the end (35), grandma would cut the sandwich in two halves diagonally—the only way to go.

Those peanut butter and banana sandwiches with a glass of milk will live on in my memory forever. You can have your sliced bananas. You can have your toasted sandwiches. You can have your crunchy peanut butter. It is 40 years since I used to eat my grandma's sandwiches, but it seems like yesterday. When I got older and had children of my own, I started making peanut butter and banana sandwiches for them just like my grandmother did for me. I've seen different styles and recipes, but I'll stick to my grandma's peanut butter and mashed banana sandwiches. They were—and still are—simply the best.

31. Choose the best replacement.
   a. NO CHANGE
   b. peanut-butter
   c. peanut butter
   d. peanutbutter,

32. Choose the best replacement sentence.
   a. NO CHANGE
   b. making peanut butter and banana sandwiches my brothers and me
   c. making me peanut butter and my brothers banana sandwiches
   d. making I and my brothers peanut butter and banana sandwiches

33. Choose the best replacement.
   a. NO CHANGE
   b. then she would carefully add
   c. then carefully adding
   d. then carefully added

34. Choose the best replacement word or phrase.
   a. NO CHANGE
   b. The next step was the best
   c. The final step was the best step
   d. Next, the final step was the best

35. Choose the best replacement word or phrase.
   a. NO CHANGE
   b. Suddenly
   c. Immediately
   d. Finally

Questions 36–40 are based on the following passage.

> Writing academic papers is a skill that develops over time. From the earliest grades in school, students learn to spell correctly. They learn how to write coherent sentences with proper grammar and punctuation, and they learn the importance of beginning, middle, and end. In higher education, these skills become more nuanced.
>
> When writing an academic paper, spelling, punctuation, and content development are key. (36) What will your paper be about? How will you introduce the subject? What information will you include in the body? How do you transition from one point to another? How will you wrap it all up so that the paper seamlessly flows from beginning to end?
>
> Transitions are an extremely important part of writing. Without proper transitions, a paper might seem choppy and disconnected. (37) The points the author is trying to make might be less easily understood, and the paper itself risks losing credibility. Transitional phrases have to be carefully planned out and strategically placed. For example, if the paper is giving a chronological order of events or step-by-step instructions, transitional phrases that indicate what order to follow will likely be used. There are transitional words and phrases that emphasize a point, demonstrating (38) an opposing point of view, present a condition, introduce examples in support of an argument, and so much more. There are transitional phrases authors use to show a consequence

or an effect following an event. There are even transitional phrases that are used to indicate a specific time or place. Transitional words act as a <u>bridge connecting</u> (39) ideas and strengthening a paper's coherency. <u>Nonetheless</u> (40), transitional words are a very important part of any academic paper, and the more skilled authors become in knowing when to use the most appropriate transitions, the stronger their papers will be.

36. Choose the best replacement phrase.
    a. NO CHANGE
    b. students should consider spelling, punctuation, and content development.
    c. spelling, punctuation, and content development, are key.
    d. the paper needs to contain proper spelling, punctuation, and convent development.

37. Choose the best replacement passage.
    a. NO CHANGE
    b. Transitions are an extremely important part of writing without which, a paper might seem choppy and disconnected.
    c. Transitions are an extremely important part of writing. Additionally, without them, a paper might seem choppy and disconnected.
    d. Transitions are an extremely important part of writing without them, a paper might seem choppy and disconnected.

38. Choose the best replacement sentence.
    a. NO CHANGE
    b. to demonstrate
    c. demonstrate
    d. for demonstrating

39. Choose the best replacement.
    a. NO CHANGE
    b. bridge of connection
    c. bridge connecting,
    d. bridge to connections

40. Choose the best replacement word or phrase.
    a. NO CHANGE
    b. Overall,
    c. Additionally,
    d. Next,

Questions 41–45 are based on the following passage.

It seems as though many people are still confused about global warming and climate change, and a lot of <u>there</u> (41) confusion stems from misinformation. There are many skeptics who think that both of these terms are synonymous, but they are not. The terms <u>"global warming" and "climate change"</u> (42) refer to two completely different phenomena. The former refers to the rising of average global temperatures, which has been trending for a long time. "Climate change," <u>on the other hand,</u> (43) refers to changes in climate around the world, like precipitation changes, and the prevalence of droughts, heat waves, and other extreme climate phenomena.

The confusion is <u>twofold.</u> (44) Firstly, since skeptics don't see steadily warmer seasons, they don't believe in global warming. <u>Additionally,</u> (45) because they believe both terms are synonymous, they actually think that scientists purposely changed the term "global warming" to "climate change," assuming they knew they were wrong about the Earth getting warmer. Nothing could be further from the truth. When people hear the term "global warming," they tend to focus on the connotation of the word "warming." Since they still feel they are experiencing the four seasons of winter, spring, summer, and fall, and since each season still goes through approximately the same phases—cold, wet, hot, and cool—many tend to believe there is no truth to the theory behind global warming. And because they hold fast to these false beliefs, they are further skeptical of the scientific evidence behind both global warming and climate change.

There are many challenges that face scientists today, and some of those challenges are in striving to educate the public on issues that greatly affect us all. The more informed the general public becomes, the better chances we have in slowing down the effects of both climate change and global warming. We need to become more conscious about what steps we can take to leave less of a carbon footprint and hopefully leave this world better than we found it—or at the very least, prevent it from becoming worse.

41. Choose the best replacement word or phrase.
    a. NO CHANGE
    b. these
    c. they're
    d. their

42. Choose the best replacement.
    a. NO CHANGE
    b. "global warming and climate change"
    c. "global warming, and climate change"
    d. 'global warming' and 'climate change'

43. Choose the best replacement word or phrase
    a. NO CHANGE
    b. however,
    c. on the whole,
    d. likewise,

44. Correct the punctuation error.
    a. NO CHANGE
    b. twofold:
    c. twofold;
    d. twofold,

Math

No Calculator Questions

1. Given the linear function $g(x) = \frac{1}{4}x - 2$, which domain value corresponds to a range value of $\frac{1}{8}$?

   a. $\frac{17}{2}$
   b. $-\frac{63}{32}$
   c. 0
   d. $\frac{2}{17}$

2. What are the zeros of the following cubic function?

$$g(x) = x^3 - 2x^2 - 9x + 18$$

   a. 2, 3
   b. 2, 3, -2
   c. 2, 3, -3,
   d. 2, -2

3. For which values of x is the following function undefined?

$$f(x) = \frac{x^2 + 3x + 1}{-3x + x^2}$$

   a. 2, 3
   b. 0, 3
   c. 1, 2
   d. 0, 1

4. If $h(x) = -6x + 11$, what is $h(-2)$?
   a. 1
   b. -1
   c. 23
   d. -23

5. For which of the following values of x is the following function undefined?

$$f(x) = \frac{-10}{x^2 - 10x + 25}$$

   a. -5
   b. 5
   c. -10
   d. 0

6. Given the following functions, what is the value of $g(2)$?

$$f(x) = -9x + 2$$
$$g(x) = f(x) - 4$$

   a. -16
   b. -20
   c. -2
   d. -4

7. Given that $m = 2\sqrt{5}$ and $3m = \sqrt{4n}$, which is the correct value of n?

a. 45

b. $\sqrt{45}$

c. 36

d. 145

8. Given the equation $6\sqrt{x-2} = 5\sqrt{x}$, what is the value of x?

a. 72

b. $\frac{72}{11}$

c. $\frac{6}{11}$

d. 44

9. Which are the values of x that satisfy the equation $2x^2 - 10x + 3 = 0$?

a. $5 \pm \frac{\sqrt{19}}{2}$

b. $5 \pm \sqrt{19}$

c. $\frac{5}{2} \pm \frac{\sqrt{19}}{2}$

d. $-\frac{5}{2} \pm \sqrt{19}$

10. What is the value of the constant b in the following equation?

$$-4x(x-9) + 8(x-3) = ax^2 + bx + c$$

a. -4

b. -44

c. -24

d. 44

11. A circle has a radius of 8 and the y-coordinate of the center in the xy-plane is -6. Two points on the circumference of the circle are (4, 2) and (-4, -6). Which of the following is the x-coordinate of the center?

a. 4

b. -4

c. 6

d. 11

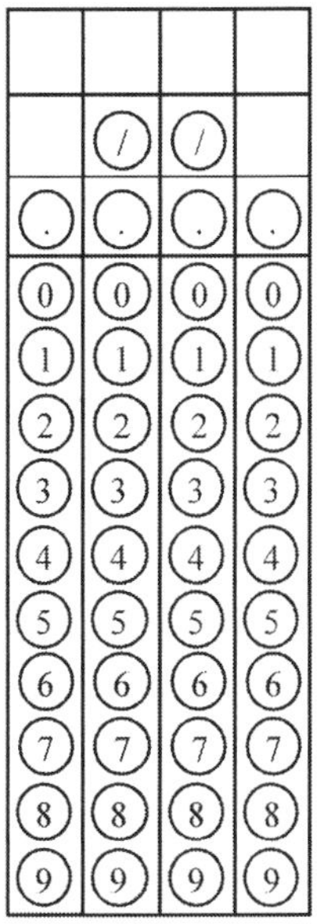

12. What is the missing length *x*?

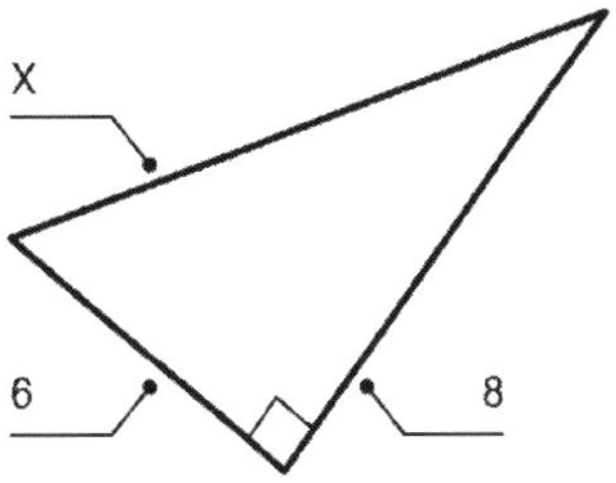

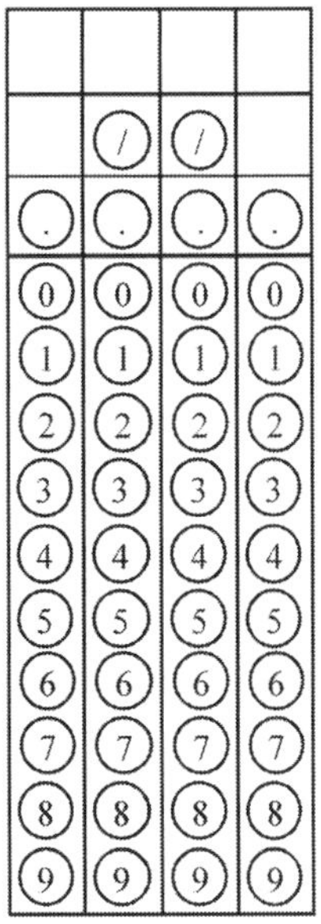

13. What is the volume of the given figure in cubic centimeters?

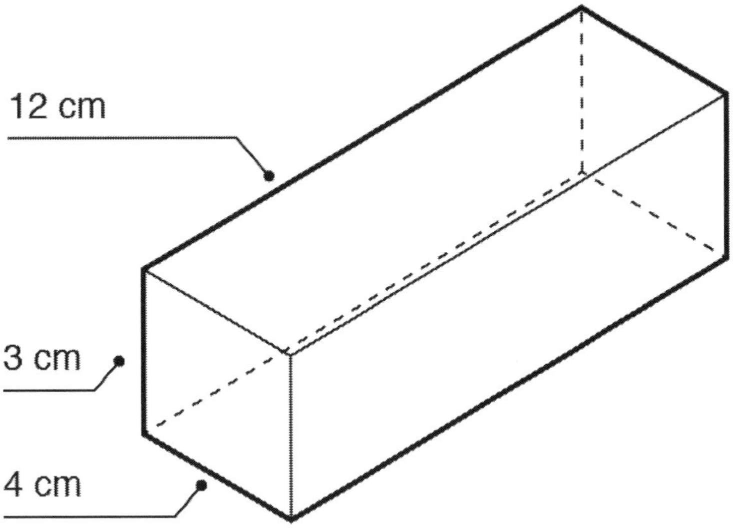

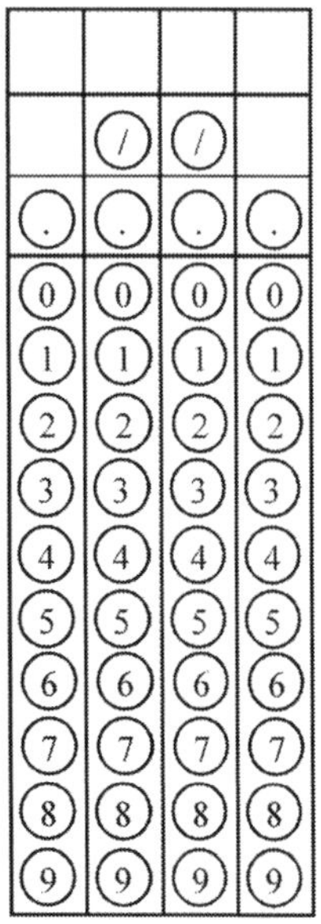

14. Given the equation $|x - 19| + 8 = 26$, what is a possible value of x?

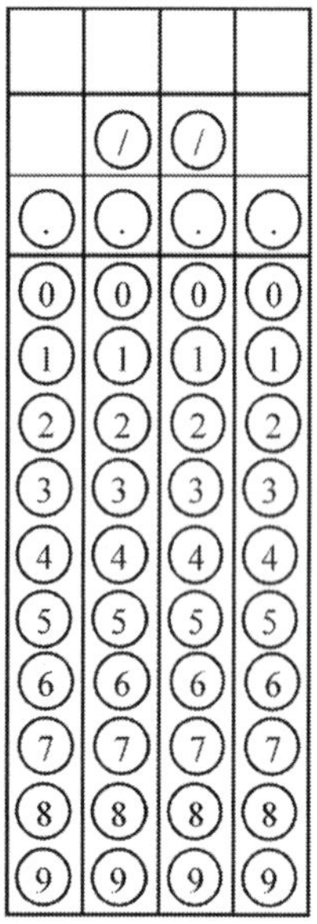

15. If $f(6) = 11$, $f(17) = 13$, $g(13) = 9$, and $g(6) = 17$, what is the value of $f(g(6))$?

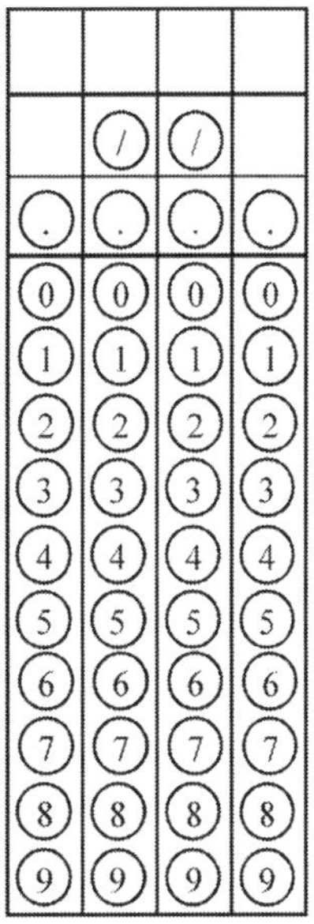

16. What is the smallest possible solution for the following equation?

$$12x^2 - 11x + 9 = 9x^2 - 3x + 4$$

Calculator Questions

17. What is the solution to the equation $3(x + 2) = 14x - 5$?
   a. $x = 1$
   b. $x = 0$
   c. All real numbers
   d. There is no solution

18. What is the solution to the equation $10 - 5x + 2 = 7x + 12 - 12x$?
   a. $x = 1$
   b. $x = 0$
   c. All real numbers
   d. There is no solution

19. Which of the following is the result when solving the equation $4(x + 5) + 6 = 2(2x + 3)$?
   a. $x = 26$
   b. $x = 6$
   c. All real numbers
   d. There is no solution

20. Two consecutive integers exist such that the sum of three times the first and two less than the second is equal to 411. What are those integers?
   a. 103 and 104
   b. 104 and 105
   c. 102 and 103
   d. 100 and 101

21. In a neighborhood, 15 out of 80 of the households have children under the age of 18. What percentage of the households have children?
   a. 0.1875%
   b. 18.75%
   c. 1.875%

d. 15%

22. If a car is purchased for $15,395 with a 7.25% sales tax, what is the total price?
    a. $15,395.07
    b. $16,511.14
    c. $16,411.13
    d. $15,402

23. Which type of graph best represents a continuous change over a period of time?
    a. Bar graph
    b. Line graph
    c. Pie graph
    d. Histogram

24. Using the graph below, what is the mean number of visitors for the first 4 hours?

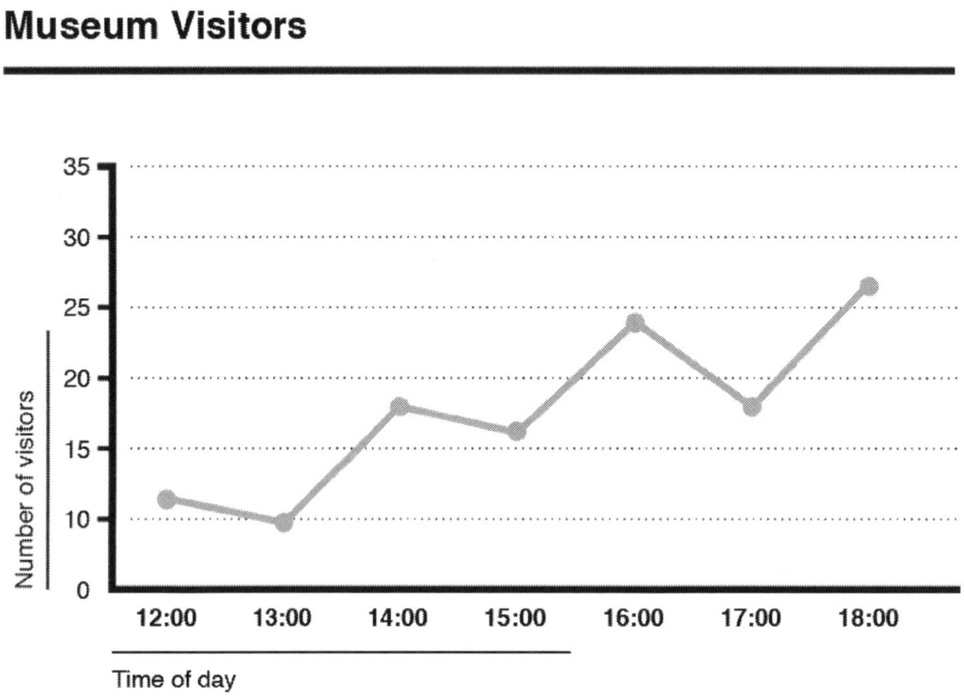

    a. 12
    b. 13
    c. 14
    d. 15

25. What is the mode for the grades shown in the chart below?

Science Grades	
Jerry	65
Bill	95
Anna	80
Beth	95
Sara	85
Ben	72
Jordan	98

a. 65
b. 33
c. 95
d. 90

26. What type of relationship is there between age and attention span as represented in the graph below?

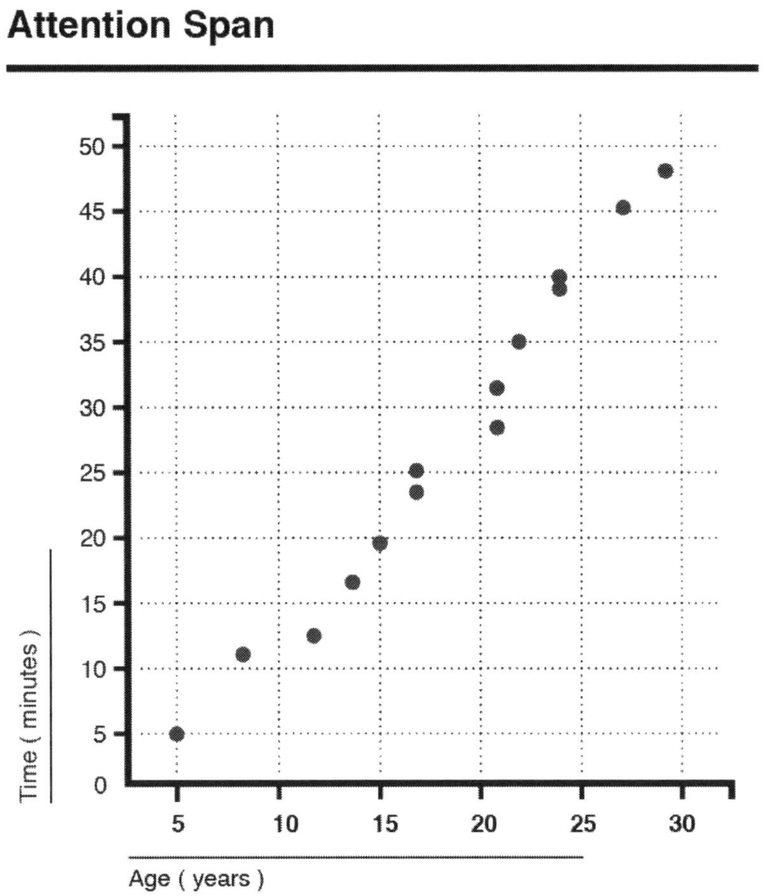

a. No correlation
b. Positive correlation
c. Negative correlation
d. Weak correlation

27. How many kiloliters are in 6 liters?
   a. 6,000
   b. 600
   c. 0.006
   d. 0.0006

28. Which of the following relations is a function?
   a. {(1, 4), (1, 3), (2, 4), (5, 6)}
   b. {(-1, -1), (-2, -2), (-3, -3), (-4, -4)}
   c. {(0, 0), (1, 0), (2, 0), (1, 1)}
   d. {(1, 0), (1, 2), (1, 3), (1, 4)}

29. Find the indicated function value: $f(5)$ for $f(x) = x^2 - 2x + 1$.
   a. 16
   b. 1
   c. 5
   d. Does not exist

30. What is the domain of $f(x) = 4x^2 + 2x - 1$?
   a. $(0, \infty)$
   b. $(-\infty, 0)$
   c. $(-\infty, \infty)$
   d. $(-1, 4)$

31. What is the range of the polynomial function $f(x) = 2x^2 + 5$?
   a. $(-\infty, \infty)$
   b. $(2, \infty)$
   c. $(0, \infty)$
   d. $[5, \infty)$

32. For which two values of x are the following functions equal?

$$f(x) = 4x + 4$$
$$g(x) = x^2 + 3x + 2$$

   a. 1, 0
   b. -2, -1
   c. -1, 2
   d. 1, 2

33. The population of coyotes in the local national forest has been declining since 2000. The population can be modeled by the function $y = -(x - 2)^2 + 1600$, where y represents number of coyotes and x represents the number of years past 2000. When will there be no more coyotes?
   a. 2020
   b. 2040
   c. 2012
   d. 2042

34. A ball is thrown up from a building that is 800 feet high. Its position s in feet above the ground is given by the function $s = -32t^2 + 90t + 800$, where t is the number of seconds since the ball was thrown. How long will it take for the ball to come back to its starting point? Round your answer to the nearest tenth of a second.

a. 0 seconds
b. 2.8 seconds
c. 3 seconds
d. 8 seconds

35. What is the domain of the following rational function?

$$f(x) = \frac{x^3 + 2x + 1}{2 - x}$$

a. $(-\infty, -2) \cup (-2, \infty)$
b. $(-\infty, 2) \cup (2, \infty)$
c. $(2, \infty)$
d. $(-2, \infty)$

36. Is the function $g(x) = 7x^3 + 5x - 2$ odd, even, both even and odd, or neither even nor odd?

a. Odd
b. Even
c. Both
d. Neither

37. A study of adult drivers finds that it is likely that an adult driver wears his seatbelt. Which of the following could be the probability that an adult driver wears his seat belt?

a. 0.90
b. 0.05
c. 0.25
d. 0

38. What is the solution to the following linear inequality?

$$7 - \frac{4}{5}x < \frac{3}{5}$$

a. $(-\infty, 8)$
b. $(8, \infty)$
c. $[8, \infty)$
d. $(-\infty, 8]$

39. What is the solution to the following system of linear equations?

$$2x + y = 14$$
$$4x + 2y = -28$$

a. (0, 0)
b. (14, -28)
c. All real numbers
d. There is no solution

40. Which of the following is perpendicular to the line $4x + 7y = 23$?

a. $y = -\frac{4}{7}x + 23$

b. $y = \frac{7}{4}x - 12$

c. $4x + 7y = 14$

d. $y = -\frac{7}{4}x + 11$

41. What is the solution to the following system of equations?

$$2x - y = 6$$
$$y = 8x$$

a. (1, 8)

b. (-1, 8)

c. (-1, -8)

d. There is no solution.

42. The mass of the moon is about 7.348×10^{22} kilograms and the mass of Earth is 5.972×10^{24} kilograms. How many times GREATER is Earth's mass than the moon's mass?

a. 8.127×10^{1}

b. 8.127

c. 812.7

d. 8.127×10^{-1}

43. The percentage of smokers above the age of 18 in 2000 was 23.2 percent. The percentage of smokers over the age of 18 in 2015 was 15.1 percent. Find the average rate of change in the percentage of smokers over the age of 18 from 2000 to 2015.

a. -.54 percent

b. -54 percent

c. -5.4 percent

d. -15 percent

44. What is the solution to the following compound inequality?

$$-14 < 4x + 6 < 18$$

a. [-5, 3]

b. [3, 5]

c. (-5, 3)

d. [-5, 3)

45. What is the equivalent exponential form of the following logarithmic function?

$$f(x) = \log_5(x + 3)$$

a. $3^y = x$

b. $5^y = x$

c. $5^{(x+3)} = y$

d. $5^y = x + 3$

46. Based on the equation $\frac{1}{4}(14x + 15) + 10 = 7x + 12$, what is the value of $6x + 7$?

a. 10

b. $\frac{1}{2}$

c. 3

d. -10

47. The total inventory of a small grocery store consists of $\frac{1}{4}$ produce items, $\frac{1}{3}$ dairy items, $\frac{1}{6}$ freezer items, and 142 items that are not one of these three types. How many different items are in the grocery store?

a. 142

b. 1704

c. 568

d. 571

48. If the volume of a sphere is the same as a cylinder that has a radius of 6 inches and a height of 10 inches, what is the radius, in inches, of the sphere?

a. 10

b. $3\sqrt[3]{10}$

c. $\sqrt[3]{30}$

d. 13

49. Olive oil is being transferred from a rectangular storage container that has dimensions 5 inches by 9 inches by 12 inches into a cylindrical storage container that has a diameter of 5 inches. What is the minimum possible length for such a cylindrical storage container that will hold all of the olive oil?

a. $\frac{432}{5\pi}$ inches

b. $\frac{108}{5\pi}$ inches

c. 6.25π inches

d. 5π inches

50. If $a + bi$ represents the result of multiplying $4.5 + 6.5i$ times $9 - 5.4i$, what is the value of a?

a. 40.5

b. 75.6

c. 34.2

d. 5.4

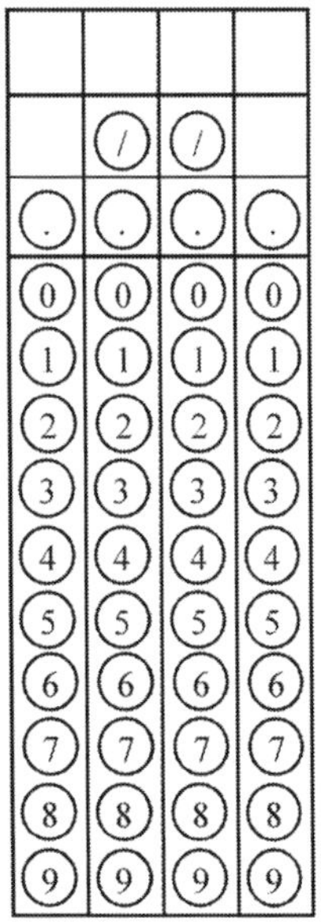

51. Given the system of equations listed below, what is the value of $3x + 4y$?

$$3x + 9y = 18$$
$$x - 3y = 15$$

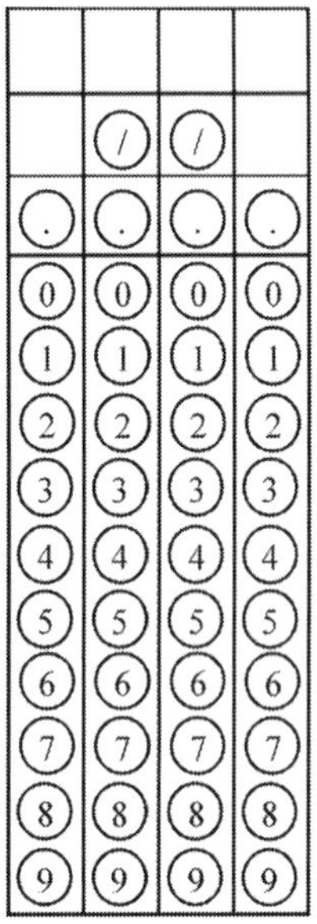

52. The equations $hy = -4.2x - 9.1$ and $8.4x + 4y = -18.2$ have infinitely many solutions. What is the value of h?

53. The equation $\frac{4}{5}x + \frac{5}{8}y = 2$ is graphed in the xy-plane. What is the y-coordinate of its y-intercept?

54. The solution to the system of equations is an ordered pair of the form (x, y). What is the value of y?

$$-x + y = 6$$

$$\frac{1}{4}(3x + 9y) = 12$$

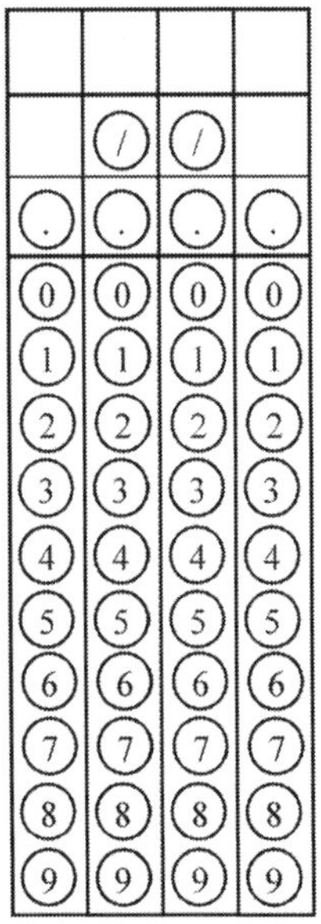

55. If $g(x) = 4x + 9$ and $g(x - c) = 4x - 22$, what is the value of c?

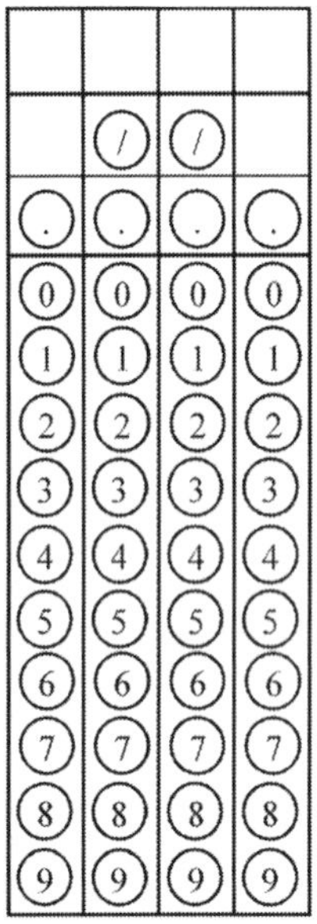

56. A shoe manufacturer models profit with the following function:

$$P(x) = 175x - x^2$$

This calculates their profit for selling 175 pairs of shoes when they charge x dollars per pair. How much more profit per pair, in dollars, will the show manufacturer earn if it charges $75 per pair versus $60 per pair? Round your answer to the nearest cent and do not use the dollar symbol in your answer.

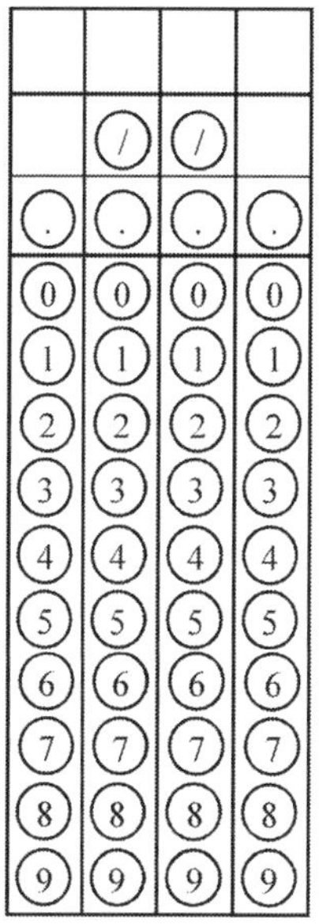

57. Given that $b^5 = -32$, what is the value of the following expression? Round your answer to the nearest thousandths place.

$$\frac{-160\left(\frac{1}{b^3}\right)}{(b^4)^2}$$

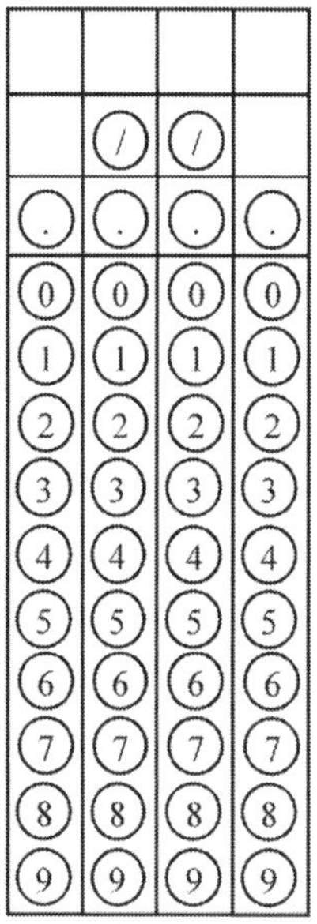

58. Jackie spends the same amount of money each month at the coffee shop. Each month, she spends half of her amount on donuts and the rest on coffee. After 8 months, she has spent \$37.50 on donuts. If after two years, she has spent \$112.50 on donuts, how much does she spend in total each month at the coffee shop? Round to the nearest cent and do not include the dollar symbol in your answer.

Writing Prompt

Read the passage below. Take into consideration how the author uses evidence to support their claims. Consider the following:

- Evidence to support claims, such as facts or examples
- Logical reasoning to develop main ideas and connect claims and evidence
- Stylistic elements, such as how figurative language appeals to the readers' emotions

Write an essay explaining how the author of the following passage builds their argument. Examine how they use one or more of the features above to enhance the logic and persuasiveness of their claim. Remember your job isn't to agree or disagree with the author, but to analyze the way in which they build their argument.

> According to the plan of the convention, all judges who may be appointed by the United States are to hold their offices *during good behavior*, which is conformable to the most approved of the State constitutions and among the rest, to that of this State. Its propriety having been drawn into question by the adversaries of that plan, is no light symptom of the rage for objection, which disorders their imaginations and judgments. The standard of good behavior for the continuance in office of the judicial magistracy, is certainly one of the most valuable of the modern improvements in the practice of government. In a monarchy it is an excellent barrier to the despotism of the prince; in a republic it is a no less excellent barrier to the encroachments and oppressions of the representative body. And it is the best expedient which can be devised in any government, to secure a steady, upright, and impartial administration of the laws.

Answer Explanations #1

Reading

1. D: Neutrality due to the style of the report. The report is mostly objective; we see very little language that entails any strong emotion whatsoever. The story is told almost as an objective documentation of a sequence of actions—we see the president sitting in his box with his wife, their enjoyment of the show, Booth's walk through the crowd to the box, and Ford's consideration of Booth's movements. There is perhaps a small amount of bias when the author mentions the president's "worthy wife." However, the word choice and style show no signs of excitement, sadness, anger, or apprehension from the author's perspective, so the best answer is Choice *D*.

2. B: Mr. Ford assumed Booth's movement throughout the theater was due to being familiar with the theater. Choice *A* is incorrect; although Booth does eventually make his way to Lincoln's box, Mr. Ford does not make this distinction in this part of the passage. Choice *C* is incorrect; although the passage mentions "companions," it mentions Lincoln's companions rather than Booth's companions. Choice *D* is incorrect; the passage mentions "dress circle," which means the first level of the theater, but this is different from a "dressing room."

3. C: A lead singer leaves their band to begin a solo career, and the band drops in sales by 50 percent on their next album. The original source of the analogy displays someone significant to an event who leaves, and then the event becomes the worst for it. We see Mr. Sothern leaving the theater company, and then the play becoming a "very dull affair." Choice *A* depicts a dancer who backs out of an event before the final performance, so this is incorrect. Choice *B* shows a basketball player leaving an event, and then the team makes it to the championship but then loses. This choice could be a contestant for the right answer; however, we don't know if the team has become the worst for his departure or the better for it. We simply do not have enough information here. Choice *D* is incorrect. The actor departs an event, but there is no assessment of the quality of the movie. It simply states what actors filled in instead.

4. A: A chronological account in a fiction novel of a woman and a man meeting for the first time. It's tempting to mark Choice *A* wrong because the genres are different. Choice *A* is a fiction text, and the original passage is not a fictional account. However, the question stem asks specifically for organizational structure. Choice *A* is a chronological structure just like the passage, so this is the correct answer. The passage does not have a cause and effect or problem/solution structure, making Choices *B* and *D* incorrect. Choice *C* is tempting because it mentions an autobiography; however, the structure of this text starts at the end and works its way toward the beginning, which is the opposite structure of the original passage.

5. C: The word *adornments* would LEAST change the meaning of the sentence because it's the most closely related word to *festoons*. The other choices don't make sense in the context of the sentence. *Feathers* of flags, *armies* of flags, and *buckets* of flags are not as accurate as the word *adornments* of flags. The passage also talks about other décor in the setting, so the word adornments fits right in with the context of the paragraph.

6. D: To recount in detail the events that led up to Abraham Lincoln's death. Choice *A* is incorrect; the author makes no claims and uses no rhetoric of persuasion towards the audience. Choice *B* is incorrect, though it's a tempting choice; the passage depicts the setting in exorbitant detail, but the setting itself is not the primary purpose of the passage. Choice *C* is incorrect; one could argue this is a narrative, and the passage is about Lincoln's last few hours, but this isn't the *best* choice. The best choice recounts the details that leads up to Lincoln's death.

7. D: The word *deleterious* can be best interpreted as referring to the word *ruinous*. The first paragraph attempts to explain the process of milk souring, so the "acid" would probably prove "ruinous" to the growth of bacteria and cause souring. Choice *A*, *amicable*, means friendly, so this does not make sense in context. Choice *B*, *smoldering*, means to boil or simmer, so this is also incorrect. Choice *C*, luminous, has positive connotations and doesn't make sense in the context of the passage. Luminous means shining or brilliant.

8. B: The author begins by explaining a process or phenomenon, then gives the history of the study of this phenomenon, then ends by presenting the effects of this phenomenon. The author explains the process of souring in the first paragraph by informing the reader that "it is due to the action of certain of the milk bacteria upon the milk sugar which converts it into lactic acid, and this acid gives the sour taste and curdles the milk." In paragraph two, we see how the phenomenon of milk souring was viewed when it was "first studied," and then we proceed to gain insight into "recent investigations" toward the end of the paragraph. Finally, the passage ends by presenting the effects of the phenomenon of milk souring. We see the milk curdling, becoming bitter, tasting soapy, turning blue, or becoming thread-like. All of the other answer choices are incorrect.

9: A: To inform the reader of the phenomenon, investigation, and consequences of milk souring. Choice *B* is incorrect because the passage states that Bacillus acidi lactici is not the only cause of milk souring. Choice *C* is incorrect because, although the author mentions the findings of researchers, the main purpose of the text does not seek to describe their accounts and findings, as we are not even told the names of any of the researchers. Choice *D* is tricky. We do see the author present us with new findings in contrast to the first cases studied by researchers. However, this information is only in the second paragraph, so it is not the primary purpose of the *entire passage*.

10. C: Milk souring is caused mostly by a species of bacteria identical to that of Bacillus acidi lactici although there are a variety of other bacteria that cause milk souring as well. Choice *A* is incorrect because it contradicts the assertion that the souring is still caused by a variety of bacteria. Choice *B* is incorrect because the ordinary cause of milk souring *is known* to current researchers. Choice *D* is incorrect because this names mostly the effects of milk souring, not the cause.

11. C: The study of milk souring has improved throughout the years, as we now understand more of what causes milk souring and what happens afterward. None of the choices here are explicitly stated, so we have to rely on our ability to make inferences. Choice *A* is incorrect because there is no indication from the author that milk researchers in the past have been incompetent—only that recent research has done a better job of studying the phenomenon of milk souring. Choice *B* is incorrect because the author refers to dairymen in relation to the effects of milk souring and their "troubles" surrounding milk souring and does not compare them to milk researchers. Choice *D* is incorrect because we are told in the second paragraph that only certain types of bacteria are able to sour milk. Choice *C* is the best answer choice here because although the author does not directly state that the study of milk souring has improved, we can see this might be true due to the comparison of old studies to newer studies, and the fact that the newer studies are being used as a reference in the passage.

12. A: The chemical change that occurs when a firework explodes. The author tells us that after milk becomes slimy, "it persists in spite of all attempts made to remedy it," which means the milk has gone through a chemical change. It has changed its state from milk to sour milk by changing its odor, color, and material. After a firework explodes, there is nothing one can do to change the substance of a firework back to its original form—the original substance is turned into sound and light. Choice *B* is incorrect because, although the rain overwatered the plant, it's possible that the plant is able to recover from this. Choice *C* is incorrect because although Mercury leaking out may be dangerous, the actual substance itself stays the same and does not alter into something else. Choice *D* is incorrect; this situation is not analogous to the alteration of a substance.

13. D: A paragraph showing the ways bacteria infiltrate milk and ways to avoid this infiltration. Choices *A*, *B*, and *C* are incorrect because these are already represented in the third, second, and first paragraphs. Choice *D* is the best answer because it follows a sort of problem/solution structure in writing.

14. A: What is the interpretation of the first amendment and its limitations? This is a central question to both passages. Choice *B* is incorrect; a quote mentions this at the end of the first passage, but this question is not found in the second passage. Choice *C* is incorrect, as the passages are not concerned with the definition of freedom of speech, but how to interpret it. Choice *D* is incorrect; this is a question for the second passage but is not found in the first passage.

15. C: A man is thrown in jail for encouraging a riot against the U.S. government for the wartime tactics although no violence ensues. The authors would most likely disagree over this answer choice. The author of Passage *A* says that "If a state may properly forbid murder or robbery or treason, it may also punish those who induce or counsel the commission of such crimes." This statement tells us that the author of Passage *A* would support throwing the man in jail for encouraging a riot, although no violence ensues. The author of Passage *B* states that "And we can with certitude declare that the First Amendment forbids the punishment of words merely for their injurious tendencies." This is the best answer choice because we are clear on each author's stance in this situation.

Choice *A* is tricky; the author of Passage *A* would definitely agree with this, but it's questionable whether the author of Passage *B* would agree with this. Violence does ensue at the capitol as a result of this man's provocation, and the author of Passage *B* states "speech should be unrestricted by censorship . . . unless it is clearly liable to cause direct . . . interference with the conduct of war." This answer is close, but it is not the *best* choice. Choice *B* is incorrect because we have no way of knowing what the authors' philosophies are in this situation. Choice *D* is incorrect because, again, we have no way of knowing what the authors would do in this situation, although it's assumed they would probably both agree with this.

16. A: Choice *A* is the best answer. To figure out the correct answer choice we must find out the relationship between Passage *A* and Passage *B*. Between the two passages, we have a general principle (freedom of speech) that is questioned on the basis of interpretation. In Choice *A*, we see that we have a general principle (right to die, or euthanasia) that is questioned on the basis of interpretation as well. Should euthanasia only include passive euthanasia, or euthanasia in any aspect? Choice *B* is incorrect because it does not question the interpretation of a principle, but rather describes the effects of two events that happened in the past involving contamination of radioactive substances. Choice *C* begins with a principle—that of labor laws during wartime—but in the second option, the interpretation isn't questioned. The second option looks at the historical precedent of labor laws in the past during wartime. Choice *D* is incorrect because the two texts disagree over the cause of something rather than the interpretation of it.

17. B: This is the best answer choice because the author is trying to demonstrate by the examples that anyone who incites a crime, despite the severity or magnitude of the crime, should be held accountable for that crime in some degree. Choice *A* is incorrect because the crimes mentioned are not being compared to each other, but they are being used to demonstrate a point. Choice *C* is incorrect because the author makes the same point using both of the examples and does not question the definition of freedom of speech but its ability to be limited. Choice *D* is incorrect because this sentiment goes against what the author has been arguing throughout the passage.

18. A: The idea that human beings are able and likely to change their mind between the utterance and execution of an event that may harm others. This idea most seriously undermines the claim because it brings into question the bad tendency of a crime and points out the difference between utterance and action in moral situations. Choice *B* is incorrect; this idea does not undermine the claim at hand but introduces an observation irrelevant to the claim. Choices *C* and *D* would most likely strengthen the argument's claim; or, they are at least supported by the author in Passage *A*.

19. D: To call upon the interpretation of freedom of speech to be already evident in the First Amendment and to offer a clear perimeter of the principle during war time. Choice *A* is incorrect; the passage calls upon no historical situations as precedent in this passage. Choice *B* is incorrect; we can infer that the author would not agree with this, because they state that "In war time, therefore, speech should be unrestricted . . . by punishment." Choice *C* is incorrect; this is more consistent with the main idea of the first passage.

20. A: The word that would least change the meaning of the sentence is *A*, grievance. Malcontent is a complaint or grievance, and in this context would be uttered in advocation of absolute freedom of speech. Choice *B*, cacophony, means a harsh noise; someone may express or "urge" a cacophony but it would be an awkward word in this context. Choice *C*, anecdote, is a short account of an amusing story. Since the word is a noun it fits grammatically inside the sentence, but anecdotes are usually thought out, and this word is considered "unthinking." Choice *D*, residual, means to be something of an outcome, or what is left behind or remaining, which does not make sense within this context.

21. B: The author says that the classical understanding of poetry dealt with its ability to be used to teach morality. Later, philosophers would define poetry by its ability to imitate life. Finally, during the renaissance, poetry was believed to be an imitative art that instilled morality in its readers. The rest of the answer choices are mixed together from this explanation in the passage. Poetry was never mentioned for use in entertainment, which makes Choice *D* incorrect. Choices *A* and *C* are incorrect for mixing up the chronological order.

22. C: That although most poetry was written as lyric, epic, or drama, the critics were most focused on the techniques of the epic and drama and their performance of structure and character. This is the best answer choice as portrayed by paragraph three. Choice *A* is incorrect because nowhere in the passage does it say rhetoric was more valued than poetry, although it did seem to have a more definitive purpose than poetry. Choice *B* is incorrect; this almost mirrors Choice *A*, but the critics were *not* focused on the lyric, as the passage indicates. Choice *D* is incorrect because the passage does not mention that the study of poetics was more pleasurable than the study of rhetoric.

23. A: To contemplate the differences between classical rhetoric and poetry and to consider their purposes in a particular culture. Choice *B* is incorrect; although changes in poetics throughout the years is mentioned, this is not the main idea of the passage. Choice *C* is incorrect; although this is partly true, that rhetoric within the education system is mentioned, the subject of poetics is left out of this answer choice. Choice *D* is incorrect; the passage makes no mention of poetics being a subset of rhetoric.

24. B: The correct answer choice is Choice *B*, instill. Choice *A*, imbibe, means to drink heavily, so this choice is incorrect. Choice *C*, implode, means to collapse inward, and does not make sense in this context. Choice *D*, inquire, means to investigate. This option is better than the other options, but it is not as accurate as *instill*.

25. B: The first paragraph presents us with definitions and examples of a particular subject. The second paragraph presents a second subject in the same way. The third paragraph offers a contrast of the two subjects. In the passage, we see the first paragraph defining rhetoric and offering examples of how the Greeks and Romans taught this subject. In paragraph two we see poetics being defined along with examples of its dynamic definition. In the third paragraph, we see the contrast between rhetoric and poetry characterized through how each of these were studied in a classical context.

26. D: The best answer is Choice *D*. As a football player, they taught me how to understand the logistics of the game, how my placement on the field affected the rest of the team, and how to run and throw with a mixture of finesse and strength. The content of rhetoric in the passage . . . "taught how to work up a case by drawing valid inferences from sound evidence, how to organize this material in the most persuasive order, and how to compose in clear and harmonious sentences." What we have here is three general principles: 1) it taught me how to understand logic and reason (drawing inferences parallels to understanding the logistics of the game), 2) taught me how to understand structure and organization (organization of material parallels to organization on the field) and 3) it taught me how to make the end product beautiful (how to compose in harmonious sentences parallels to how to run with finesse and strength). Each part parallels by logic, organization, then style.

27. A: Treatises is most closely related to the word *commentary*. Choice *B* does not make sense because thesauruses and encyclopedias are not written about one single subject. Choice *C* is incorrect; sermons are usually given by religious leaders as advice or teachings. Choice *D* is incorrect; anthems are songs and do not fit within the context of this sentence.

28. D: A city whose population is made up of people who seek quick fortunes rather than building a solid business foundation. Choice *A* is a characteristic of Portland, but not that of a boom city. Choice *B* is close—a boom city is one that becomes quickly populated, but it is not necessarily always populated by residents from the east coast. Choice *C* is incorrect because a boom city is not one that catches fire frequently, but one made up of people who are looking to make quick fortunes from the resources provided on the land.

29. D: Choice *D* is the best answer because of the surrounding context. We can see that the fact that Portland is a "boom city" means that the "floating class" go through, a group of people who only have temporary roots put down. This would cause the main focus of the city to be on employment and industry, rather than society and culture. Choice *A* is incorrect, as we are not told about the inhabitants being social or antisocial. Choice *B* is incorrect because the text does not talk about the culture in the East regarding European influence. Finally, Choice *C* is incorrect; this is an assumption that has no evidence in the text to back it up.

30. B: The author would say that it has as much culture as the cities in the East. The author says that Portland has "as fine churches, as complete a system of schools, as fine residences, as great a love of music and art, as can be found at any city of the East of equal size," which proves that the culture is similar in this particular city to the cities in the East.

31. A: Approximately 240,000. We know from the image that San Francisco has around 300,000 inhabitants at this time. From the text (and from the graph) we can see that Portland has 60,000 inhabitants. Subtract these two numbers to come up with 240,000.

32. C: This question requires close attention to the passage. Choice *A* can be found where the passage says "no less than six named and several unnamed varieties of the peach have thus produced several varieties of nectarine," so this choice is incorrect. Choice *B* can be found where the passage says "it is highly improbable that all these peach-trees . . . are hybrids from the peach and nectarine." Choice *D* is incorrect because we see in the passage that "the production of peaches from nectarines, either by seeds or buds, may perhaps be considered as a case of reversion." Choice *C* is the correct answer because the word "unremarkable" should be changed to "remarkable" in order for it to be consistent with the details of the passage.

33. A: The word *multiplied* is synonymous with the word *propagated*, making Choice *A* correct. Choice *B* is incorrect because *diminished* means to decrease or recede and is the opposite of *propagated*. Choice *C* is incorrect; *watered* is closed, because it pertains to the growth of trees, but it is not exactly the same thing as *propagated*. Finally, Choice *D* is incorrect; *uprooted* could also pertain to trees, but this answer is incorrect.

34. B: The author's tone in this passage can be considered objective. An objective tone means that the author is open-minded and detached about the subject. Most scientific articles are objective. Choices *A*, *C*, and *D* are incorrect. The author is not very enthusiastic on the paper; the author is not critical, but rather interested in the topic. The author is not desperate in any way here.

35. B: Choice *B* is the correct answer because the meaning holds true even if the words have been switched out or rearranged some. Choice *A* is incorrect because it has trees either bearing peaches or nectarines, and the trees in the original phrase bear both. Choice *C* is incorrect because the statement does not say these trees are "indifferent to bud-variation," but that they have "indifferently [bore] peaches or nectarines." Choice *D* is incorrect; the statement may use some of the same words, but the meaning is skewed in this sentence.

36. D: *Congealed* in this context most nearly means *thickened*, because we see liquid turning into ice. Choice *B*, *loosened*, is the opposite of the correct answer. Choices *A* and *C*, burst and shrank, are also incorrect.

37. C: Choice *A* is incorrect. We cannot infer that the passage takes place during the night time. While we do have a statement that says that the darkness thickened, this is the only evidence we have. The darkness could be thickening because it is foggy outside. We don't have enough proof to infer this otherwise. We *can* infer that the season of this narrative is in the winter time. Some of the evidence here is that "the cold became intense," and people were decorating their shops with "holly sprigs," a Christmas tradition. It also mentions that it's Christmas time at the end of the passage. Choice *B* is incorrect; we *can* infer that the narrative is located in a bustling city street by the actions in the story. People are running around trying to sell things, the atmosphere is busy, there is a church tolling the hours, etc. The scene switches to the Mayor's house at the end of the passage, but the answer says "majority," so this is still incorrect. Choice *D* is incorrect; we *can* infer that the Lord Mayor is wealthy—he lives in the "Mansion House" and has fifty cooks.

38. D: The passage tells us that the poulterers' and grocers' trades were "a glorious pageant, with which it was next to impossible to believe that such dull principles as bargain and sale had anything to do," which means they set up their shops to be entertaining public spectacles in order to increase sales. Choice *A* is incorrect; although the word "joke" is used, it is meant to be used as a source of amusement rather than something made in poor quality. Choice *B* is incorrect; that they put on a "pageant" is figurative for the public spectacle they made with their shops, not a literal play. Finally, Choice *C* is incorrect, as this is not mentioned anywhere in the passage.

39. B: The author, at least in the last few paragraphs, does not attempt to evoke empathy for the character of Scrooge. We see Scrooge lashing out at an innocent, cold boy, with no sign of affection or feeling for his harsh conditions. We see Choice *A* when the author talks about Saint Dunstan. We see Choice *C* providing a palpable experience, especially with the "piercing, searching, biting cold," among other statements. Finally, we see Choice *D* when Scrooge chases the young boy away.

40. C: The section that is talked about in the text is blazing the trail, Choice *C*. The passage states that one must blaze the trail by "bending down and breaking branches of trees, underbrush, and bushes." The girl in the image is bending a branch in order to break it so that she won't get lost by "blazing the trail."

41. B: Choice *B* is the best answer here; the sentence states "In unknown regions take a responsible guide with you, unless the trail is short, easily followed, and a frequented one." Choice *A* is incorrect; the passage does not state that you should try and explore unknown regions. Choice *C* is incorrect; the passage talks about trails that contain pitfalls, traps, and boggy places, but it does not say that *all* unknown regions contain these things. Choice *D* is incorrect; the passage mentions "rail" and "boat" as means of transport at the beginning, but it does not suggest it is better to travel unknown regions by rail.

42. D: Choice *D* is correct; it may be real advice an experienced hiker would give to an inexperienced hiker. However, the question asks about details in the passage, and this is not in the passage. Choice *A* is incorrect; we do see the author encouraging the reader to learn about the trail beforehand . . . "wet or dry; where it leads; and its length." Choice *B* is also incorrect, because we do see the author telling us the time will lengthen with boggy or rugged places opposed to smooth places. Choice *C* is incorrect; at the end of the passage, the author tells us "do not go alone through lonely places . . . unless you are quite familiar with the country and the ways of the wild."

43. C: The best answer here is Choice *C*, "Follow the false trail back to its beginning so that you can rediscover the real trail." Choices *A* and *D* are represented in the text; but this is advice on how to blaze a trail, not what to do when being led astray by a false trail. Choice *B* is incorrect; this is the opposite of what the text suggests doing.

44. A: The author is doing translation work. We see this very clearly in the way the author talks about staying truthful to the original language of the text. The text also mentions "translation" towards the end. Criticism is taking an original work and analyzing it, making Choice *B* incorrect. The work is not being tested for historical validity, but being translated into the English language, making Choice *C* incorrect. The author is not writing a biography, as there is nothing in here about Pushkin himself, only his work, making Choice *D* incorrect.

45. D: You would most likely find this in the preface. A preface to a text usually explains what the author has done or aims to do with the work. An appendix is usually found at the end of a text and does not talk about what the author intends to do to the work, making Choice *A* incorrect. A table of contents does not contain prose, but bullet points listing chapters and sections found in the text, making Choice *B* incorrect. Choice *C* is incorrect; the first chapter would include the translation work (here, poetry), and not the author's intentions.

46. B: To retain the truth of the work. The author says that "music and rhythm and harmony are indeed fine things, but truth is finer still," which means that the author stuck to a literal translation instead of changing up any words that might make the English language translation sound better.

47. A: The author is trying to say that he is not regretful for any crudeness on the part of the text because the translation was done to the best of the author's ability. The author asks for "no forgiveness" because what was done was the best the author could do with the text following a literal translation. Choice *B* is incorrect; the author does *not* ask for forgiveness. Choice *C* is incorrect; the author does mention God, but as a rhetorical device to make the translation seem an act of service rather than a hobby. Choice *D* is incorrect; no fighting is mentioned in this paragraph.

48. D: The author does not consider modern architecture to be "cathedral-crowned grim and dim castles." The author is speaking of feudal architecture when they say this phrase. Choices *A, B,* and *C* are all mentioned in the text as characteristics of modern architecture, especially in the cities of Chicago and Pittsburgh.

49. C: Architecture reflects—in shape, color, and form—the attitude of the society which built it. Choice *A* is too specific and is taken too literal. It is not that the architecture represents the builders; it represents the builder's culture. Choice *B* is also incorrect; we do see the words "at its worst as at its best," but the meaning of this is skewed in Choice *B*. Choice *D* is incorrect; the statement does not suggest this analysis.

50. C: The author refers to the "mediaeval man" in order to make the audience realize the irony produced from pitying him. The author says that we pity him for the dirt he lived in; however, we inhale smoke and dirt from our own skies. Choices *A, B,* and *D* are incorrect, as they just miss the mark of the statement.

51. B: The text talks about castles and cathedrals as examples of architecture from medieval Europe. Factories and skyscrapers are considered pre-modern or modern architecture. Canyons are not buildings, so Choice *A* is incorrect.

52. B: The author's tone is best described as "aggrieved" because it is annoyed, offended, and disgruntled by the "power, strangely mingled with timidity," the "ingenuity, frequently misdirected" aspects of modern architecture. The author calls out its "giant energy" and "grimy and dreariness," along with its "grim, dim canyons of our city streets."

Writing and Language

1. A: Choice *A* is the best answer. To write a direct quotation, double quotation marks are placed at the beginning and ending of the quoted words. All punctuation specifically related to the quoted phrase or sentence is kept inside the quotation marks. Single quotation marks, as used in Choices *B* and *C*, are not used for dialogue in American English.

2. B: The best answer is *B*. The reciprocal pronoun "one another" generally refers to more than two people.

3. C: The best answer is *C*. "Lulu brought this kitten home only six weeks ago, and she was so fond of her." There are two independent clauses in one sentence, so separating the two clauses with a comma is the best option. The other sentences have incorrect comma placement. Additionally, the information that Lulu was only five years old is not necessary in this particular sentence—it is out of context.

4. C: Choice *C* is the best answer. We would use the preposition "through," when talking about staying with someone "through the night."

5. C: Choice *C* is the best answer here: "Lulu continued to play so that she wouldn't scare the kitten away." Usually the words "in order" are used with the word "to." Choice *A* is incorrect because it uses the word "that" together with "to." For Choices *B* and *D*, there is no need for a comma after "so that."

6. D: The best answer is *D*: "without a set list of chores, and without ensuring the chores are completed, our household would be utterly chaotic." In this sentence, there is an interrupting phrase that must be set apart by two commas on either side of it.

7. A: Choice *A* is the best answer, no change, because "13-year-old" is correct. When an age is describing a noun and precedes the noun, hyphenation is required.

8. B. Choice *B* is the best answer, "approximately once a week." Choice *A* is not the best answer, because the adverb "approximately" interrupts the phrase "once a week," producing the awkward if grammatically correct phrase "once, approximately, a week." Choice *C* is not correct, because "bi-weekly" means twice a week or every two weeks. Choice *D* is not the best answer, because it changes the meaning to a firm "once a week."

9. B: The best answer is Choice *B*: "Both my husband and I drive our daughter." "I" is a subject pronoun, and in this sentence "I" is part of a compound subject, "my husband and I." A helpful way to answer this question is to take out the phrase "my husband" and read the sentence with each choice. Choice *A*, "myself drive our daughter" is incorrect, and so is Choice *D*, "myself, drive our daughter." Choice *C*, "me drive our daughter" is also incorrect. "I drive our daughter" is the best answer.

10. B: The best answer is Choice *B*: "who do you think has the most chores?" Since the sentence is comparing three people's chores, it is necessary to use a superlative adjective. If there were only two people, it would be correct to use the comparative adjective "more."

11. C: Choice *C* is the best answer. The words "accept" and "except" are often confused. There is no need to use the adverb "exceptionally" here, so Choice *B* is incorrect. Choice *C* is incorrect, because "expect" is a verb with a different meaning than "except."

12. B: The best answer is *B*. The word "birds" doesn't necessarily evoke a sound, but any number of actions birds make will create sound. It is best to use a qualifier here, to indicate which bird sound the person hears.

13. C: The best answer is *C*. "Ever so" is a common phrase in English. The other options are incorrectly spelled.

14. D: Choice *D* is the best answer: "you smile for the rest of the day." As the sentence is currently written, there is a dangling modifier. The sentence implies that the day, and not the person, is feeling optimistic.

15. B: The best choice is *B*. The sentence is composed of two independent clauses, and therefore requires a comma before the conjunction.

16. B: The best answer is *B*: "and I'm always the last one to class." Choice *C* changes the meaning, and Choice *D* uses an unnecessary comma. The original phrase is awkward; it is not parallel with the previous instance of the phrase "the last one." Additionally, English speakers would never omit the article "the" from the phrase "the last one."

17. C: The best answer is *C*: "I was always teased by my older sisters." The sentence in the passage demonstrates a common mistake with dangling participles. The sentence starts with the participial phrase

"Having also been the last of all the siblings." That phrase can only modify the speaker, the youngest sibling. But in the original passage, the participial phrase incorrectly modifies "my older sisters." Choice *C* is the only answer that corrects this mistake.

18. D: Choice *D* is the best answer: "Last but not least!" In American English, double quotation marks are used at the beginning and ending of a quotation, and all punctuation relating specifically to the quotation must remain inside the quotation marks. There is also no need for a comma in a simple sentence containing one simple thought, making Choice *C* incorrect.

19. B: The best answer is *B*, the semicolon. The way ideas are connected is important. In this case, the second independent clause gives an example of the first. The author says she has been dishonest; the next clause shows the instance of her dishonesty. The word "and" is not grammatically wrong, but it suggests the two facts are on the same level. Choices *C* and *D* suggest different relationships between the two facts, and these relationships do not make the most sense.

20. A: Choice *A* is the best answer. The phrase "after all" means "despite earlier problems or doubts," which is the perfect transition here because the speaker is changing their mind about being the last sibling. "Therefore" denotes a conclusion, and "however" and "on the contrary" denote a counter, and the speaker is not concluding or countering here.

21. C: The best answer is *C*, grandparents'. To show possession of a plural noun that ends in an "s," the apostrophe is placed directly after the final "s."

22. A: The original choice is the best answer here. Sauntering is not something Angela did just one time; she did it often. When referring to repeated past actions, the word "would" is used with the past tense verb, in this case "sauntered."

23. A: The best answer is *A*. There is no need to use any uppercase letters in the name because the specific name of a contest is not being used. The words in a proper noun would be capitalized.

24. C: Choice *C* is the best answer: "didn't." The entire story is written in the past tense, and the verbs must reflect this tense in order to keep the passage coherent.

25. B: The best answer is *B*: "well liked." The phrase "well liked" when it appears before the verb: "The well-liked hostess sauntered around the room." In this sentence, "well liked" appears after the verb, so no hyphen is needed.

26. B: Choice *B* is the best answer: "Jones's class." The teacher's name is Mr. Jones, and since the name "Jones" ends in an "s," in order to show possession, it is necessary to place the apostrophe directly after the "s" in the name "Jones," followed by the possessive "s."

27. B: The best answer is *B*: "He would listen to his parents' kitchen radio for hours, playing their records and singing away to all the classics." In order to keep the verb tense consistent and maintain parallel structure, the verb form "playing" should match the verb form "singing."

28. C: The best answer is *C*: "However, what he really wanted . . ." In the choices provided, "however" as a transitional word followed by a comma is the most appropriate choice because it provides a contrast inherent in the context: that he loved music but wanted to create his own.

29. D: Choice *D* is the best answer. The two actions occur one after the other, in a sequence; first Matt forgot, then he began to play. In the original sentence, the verb "had begun" suggests Matt somehow started playing before he forgot about his test.

30. D: The best answer is *D*. This sentence is set up in the past conditional, starting with the condition "if." Therefore, the use of the conditional "would fail" is necessary.

31. C: The best answer is *C*. "Peanut butter" is spelled as two separate words without a hyphen.

32. A: The best answer is *A*. The original sentence is correct. There is a joke in which a boy says to a genie, "Make me a peanut butter sandwich." The genie interprets the sentence incorrectly, and the boy is turned into a sandwich. However, there is nothing grammatically wrong with the way the boy expresses his wish, and there is nothing wrong with the original sentence. The word "sandwiches" is the direct object of the verb "make"; sandwiches are what the grandmother makes. The words "me and my brothers" are indirect objects of the verb "make"; the grandmother makes sandwiches for them. Choice *D* is incorrect, because the pronoun "I" is a subject, and the sentence requires the indirect object "me."

33. B: The best answer is *B*. To refer to repeated past actions as in this case, the word "would" is used, and in an independent clause like this one, the pronoun "she" needs to be repeated.

34. B: Choice *B* is the best answer, because it is the most coherent. The step listed in the passage is actually *not* the final step, so the word "final" is out of place. All the other choices are confusing, because they include the word "final."

35. D: The best answer is *D*. This truly is the final step in the process, when the grandmother would cut the sandwiches. "Finally" is a terminal transitional word. The original phrase "in the end" is more appropriate for an action that happened one time. The author is describing a process the grandmother did repeatedly, so "finally" is a better choice than "in the end."

36. B: Choice *B* is the best answer: "students should consider spelling, punctuation, and content development." This is an example of a dangling modifier. The writing of the paper is not done by the spelling, punctuation, or content development. The writing is done by the students, so the clause has to use "students" as the subject.

37. A: The best answer is *A*. Choice *B* is incorrect because there is a comma after "without which," and this comma interrupts the sentence. Choice *C* is incorrect because of the superfluous word "additionally." Additionally is not needed here because the second sentence acts as an explanation to the first. Choice *D* is incorrect because it is a run-on sentence; there should be a semicolon or period after "writing" and before "without them."

38. C: The best answer is *C*. Each verb in the relative clause should show parallelism with the use of the third-person singular present tense. The clause is about words that demonstrate, words that emphasize, words that present, and so on. Therefore "demonstrate" should be parallel with the verbs "emphasize," "present," and "introduce." Choices *A*, *C*, and *D* are not parallel.

39. B: The best answer is *A*. The gerund form of the verb "to connect" is "connecting." Using this form of the verb creates a parallel structure, because the sentence also uses the gerund form of "to strengthen," which is "strengthening."

40. B: The best answer is *B*. None of the other transitional words make sense. Choice *C*, "additionally," is not a good answer because the sentence is not adding a new point. Likewise, Choice *D*, "next," is not appropriate because the sentence is not about something that comes next in a sequence. Because this is the final summation of the argument, the transitional word "overall" is the best answer.

41. D: Choice *D* is the best answer. "Their" is the possessive form of the pronoun "they" and is the only coherent choice. Choices *A* and *D* are homophones; "there" and "they're" sound like "their," but they have different meanings.

42. A: The best answer is *A*. No change is needed here, because the two terms being introduced are separate and distinct, and therefore need to be separated by their own sets of double quotation marks. Choice *D* is incorrect because it uses single quotation marks.

43. B: The best answer is *B*. "On the other hand" refers to the opposite point of view from the previous one discussed, or the opposite in meaning. Climate change is not the opposite of global warming, although there are differences. "However" is the best transitional word to use here.

44. B: Choice *B* is the best answer. In this case, the colon is being used to introduce several sentences that are examples of the twofold confusion.

Math

1. A: The range value is given, and this is the output of the function. Therefore, the function must be set equal to $\frac{1}{8}$ and solved for x. Thus, $\frac{1}{8} = \frac{1}{4}x - 2$ needs to be solved. The fractions can be cleared by multiplying by the LCD 8. This results in $1 = 2x - 16$. Add 16 to both sides and divide by 2 to obtain $x = \frac{17}{2}$.

2. C: To find the zeros, set the function equal to 0 and factor the polynomial. Because there are four terms, it should be factored by grouping. Factor a common factor out of the first set of two terms, and then find a shared binomial factor in the second set of two terms. This results in:

$$x^2(x-2) - 9(x-2) = 0$$

The binomial can then be factored out of each set to get:

$$(x^2 - 9)(x - 2) = 0$$

This can be factored further as:

$$(x+3)(x-3)(x-2) = 0$$

Setting each factor equal to zero and solving results in the three zeros -3, 3, and 2.

3. B: For rational functions, the denominator can never be equal to 0. Finding the undefined values for a function is the same as finding the zeros of the expression in the denominator. To find the undefined values, set the denominator equal to 0 and solve for x. This results in $-3x + x^2 = 0$, and its solutions are $x = 0$ and $x = 3$. Therefore, the undefined values of x are 0 and 3; Choice *B* is correct.

4. C: If $h(x) = -6x + 11$, then $h(-2)$ can be found by substituting -2 in for x in the function. Therefore:

$$h(-2) = -6(-2) + 11 = 12 + 11 = 23$$

5. B: A rational function is undefined when its denominator is equal to 0. Plugging 5 into the denominator results in:

$$5^2 - 10(5) + 25 = 25 - 50 + 25 = 0$$

The other options give non-zero results. Therefore, the function is undefined when $x = 5$.

6. B: To solve, substitute $-9x + 2$ in place of $f(x)$ for the function $g(2)$ and replace x with 2:

$$g(2) = f(2) - 4$$

$$g(2) = -9(2) + 2 = -16$$

Therefore:

$$g(2) = -16 - 4 = -20$$

7. A: Given that $m = 2\sqrt{5}$ and $3m = \sqrt{4n}$, we plug $2\sqrt{5}$ in for m in the second equation to obtain:

$$6\sqrt{5} = \sqrt{4n}$$

Squaring both sides results in:

$$36(5) = 4n$$

Dividing both sides by 4 gives:

$$n = \frac{36(5)}{4} = 9(5) = 45$$

8. B: Squaring both sides results in the equation:

$$36(x - 2) = 25x$$

This is equivalent to:

$$36x - 72 = 25x$$

Adding 72 to both sides and subtracting $25x$ from both sides results in:

$$11x = 72$$

Dividing by 11 gives $x = \frac{72}{11}$. Because this is a radical equation, the answer must be checked. Plugging the result, $x = \frac{72}{11}$, into the original equation creates a true equation. Therefore, it is the solution to the original equation.

9. C: To solve this quadratic equation, the quadratic formula with $a = 2$, $b = -10$, and $c = 3$ should be used. Therefore:

$$x = \frac{-(-10) \pm \sqrt{(-10)^2 - 4(2)(3)}}{2(2)} = \frac{10 \pm \sqrt{76}}{4}$$

$$x = \frac{10 \pm 2\sqrt{19}}{4} = \frac{5}{2} \pm \frac{\sqrt{19}}{2}$$

10. D: Distribute the expressions in front of the parentheses to obtain:

$$-4x^2 + 36x + 8x - 24$$

Collecting like terms results in:

$$-4x^2 + 44x - 24$$

The coefficient on the x term is 44, so $b = 44$.

11. A: The equation for a circle with radius r and center (h, k) is:

$$(x - h)^2 + (y - k)^2 = r^2$$

From the information given, the equation is

$$(x - h)^2 + (y + 6)^2 = 64$$

Plugging in the point (4, 2) results in:

$$(4 - h)^2 + 64 = 64$$

Therefore:

$$(4 - h)^2 = 0$$

$$h = 4$$

The x-coordinate of the center of this circle is 4.

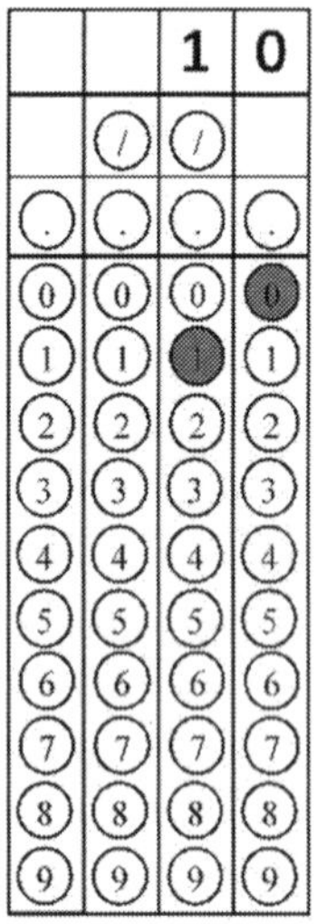

12. The Pythagorean Theorem can be used to find the missing length x because it is a right triangle. The theorem states that $6^2 + 8^2 = x^2$, which simplifies into $100 = x^2$. Taking the positive square root of both sides results in the missing value $x = 10$.

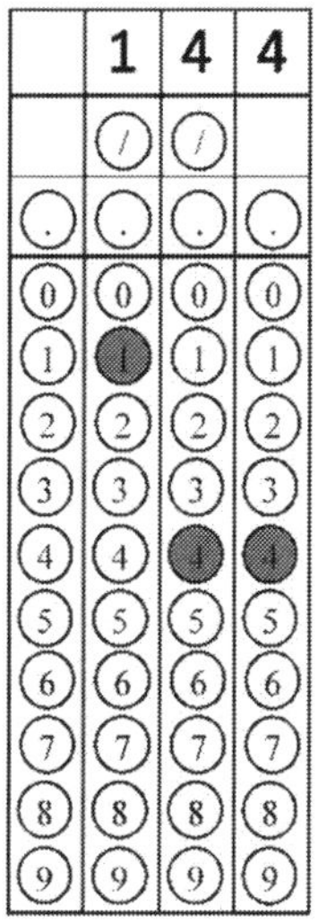

13. The volume of a rectangular prism is found by multiplying the length by the width by the height. This formula yields an answer of 144 cubic units.

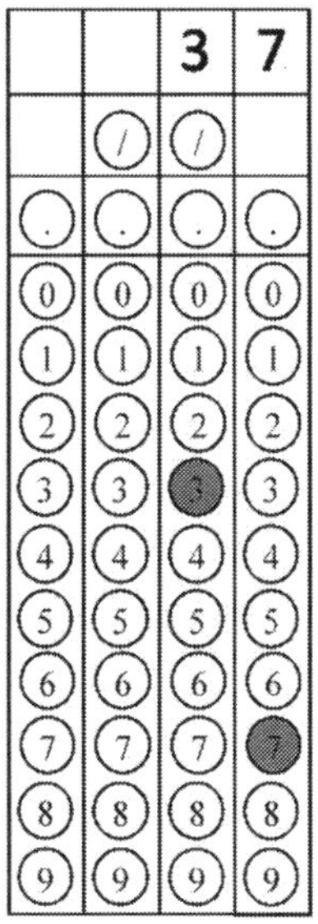

14. 1 or **37:** Starting with the equation $|x - 19| + 8 = 26$, first isolate the absolute value expression on the left side of the equals sign by subtracting 8 from both sides. Therefore, $|x - 19| = 18$. Using the definition of absolute value, either $x - 19 = 18$ or $x - 19 = -18$. There are two possible solutions: $x = 37$ or $x = 1$.

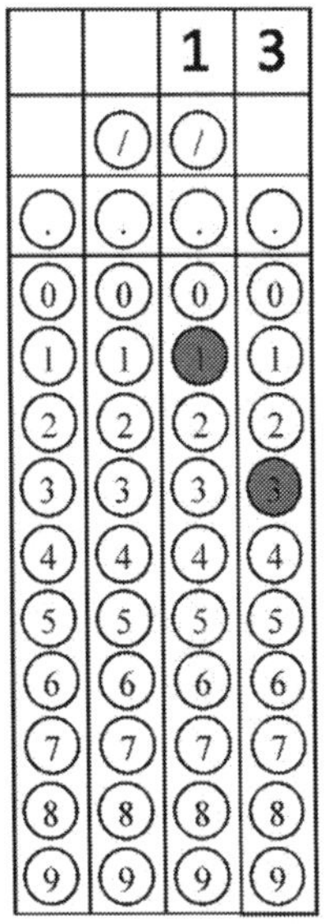

15. 13: Because $g(6) = 17$, $f(g(6)) = f(17) = 13$.

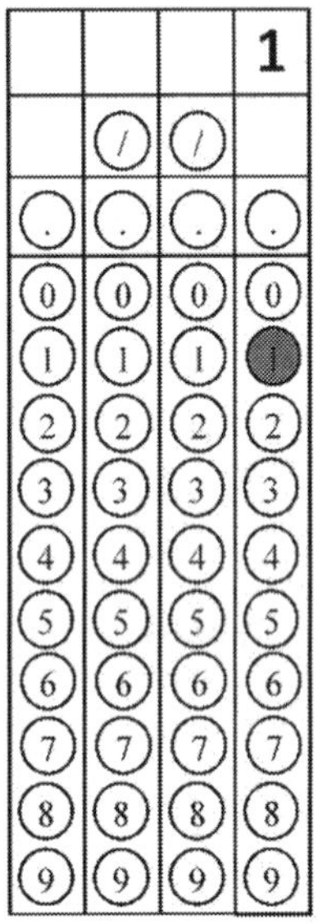

16. 1: To solve this quadratic equation, move the terms on the right-hand side of the equals sign to the left-hand side, resulting in:

$$3x^2 - 8x + 5 = 0$$

Then, use the quadratic formula with $a = 3$, $b = -8$, and $c = 5$ to obtain the solutions:

$$x = \frac{-(-8) \pm \sqrt{(-8)^2 - 4(3)(5)}}{2(3)} = \frac{8 \pm \sqrt{4}}{6} = \frac{8 \pm 2}{6}$$

Therefore, the solutions are $\frac{5}{3}$ and 1. The smallest solution is 1.

17. A: First, the distributive property must be used on the left side. This results in:

$$3x + 6 = 14x - 5$$

The addition property is then used to add 5 to both sides, and then to subtract $3x$ from both sides, resulting in $11 = 11x$. Finally, the multiplication property is used to divide each side by 11. Therefore, $x = 1$ is the solution.

18. C: First, like terms are collected to obtain:

$$12 - 5x = -5x + 12$$

Then, if the addition principle is used to move the terms with the variable, $5x$ is added to both sides and the mathematical statement $12 = 12$ is obtained. This is always true; therefore, all real numbers satisfy the original equation.

19. D: The distributive property is used on both sides to obtain:

$$4x + 20 + 6 = 4x + 6$$

Then, like terms are collected on the left, resulting in:

$$4x + 26 = 4x + 6$$

Next, the addition principle is used to subtract $4x$ from both sides, and this results in the false statement $26 = 6$. Therefore, there is no solution.

20. A: First, the variables have to be defined. Let x be the first integer; therefore, $x + 1$ is the second integer. This is a two-step problem. The sum of three times the first and two less than the second is translated into the following expression:

$$3x + (x + 1 - 2)$$

Set this expression equal to 411 to obtain:

$$3x + (x + 1 - 2) = 411$$

The left-hand side is simplified to obtain:

$$4x - 1 = 411$$

The addition and multiplication properties are used to solve for x. First, add 1 to both sides and then divide both sides by 4 to obtain $x = 103$. The next consecutive integer is 104.

21. B: First, the information is translated into the ratio $\frac{15}{80}$. To find the percentage, translate this fraction into a decimal by dividing 15 by 80. The corresponding decimal is 0.1875. Move the decimal point two places to the right to obtain the percentage 18.75%.

22. B: If sales tax is 7.25%, the price of the car must be multiplied by 1.0725 to account for the additional sales tax. Therefore:

$$15{,}395 \times 1.0725 = 16{,}511.1375$$

This amount is rounded to the nearest cent, which is $16,511.14.

23. B: A line graph represents continuous change over time. The line on the graph is continuous and not broken, as on a scatter plot. A bar graph may show change but isn't necessarily continuous over time. A pie graph is better for representing percentages of a whole. Histograms are best used in grouping sets of data in bins to show the frequency of a certain variable.

24. C: The mean for the number of visitors during the first 4 hours is 14. The mean is found by calculating the average for the four hours. Adding up the total number of visitors during those hours gives:

$$12 + 10 + 18 + 16 = 56$$

Then $56 \div 4 = 14$.

25. C: The mode for a set of data is the value that occurs the most. The grade that appears the most is 95. It's the only value that repeats in the set.

26. B: The relationship between age and time for attention span is a positive correlation because the general trend for the data is up and to the right. As age increases, so does attention span.

27. C: There are 0.006 kiloliters in 6 liters because 1 liter is 0.001 kiloliters. The conversion comes from the metric prefix -kilo which has a value of 1000. Thus, 1 kiloliter is 1000 liters, and 1 liter is 0.001 kiloliters.

28. B: The only relation in which every *x*-value corresponds to exactly one *y*-value is the relation given in Choice *B*, making it a function. The other relations have the same first component paired up to different second components, which goes against the definition of a function.

29. A: To find a function value, plug in the number given for the variable and evaluate the expression, using the order of operations (parentheses, exponents, multiplication, division, addition, subtraction). The function given is a polynomial function:

$$f(5) = 5^2 - 2 \times 5 + 1 = 25 - 10 + 1 = 16$$

30. C: The function given is a polynomial function. Anything can be plugged into a polynomial function to get an output. Therefore, its domain is all real numbers, which is expressed in interval notation as $(-\infty, \infty)$.

31. D: This is a parabola that opens up, as the coefficient on the x^2 term is positive. The smallest number in its range occurs when plugging 0 into the function $f(0) = 5$. Any other output is a number larger than 5, even when a positive number is plugged in. When a negative number gets plugged into the function, the output is positive, and same with a positive number. Therefore, the domain is written as $[5, \infty)$ in interval notation.

32. C: First, set the functions equal to one another, resulting in:

$$x^2 + 3x + 2 = 4x + 4$$

This is a quadratic equation, so the best way to find the answer is to convert it to standard form by subtracting $4x + 4$ from both sides of the equation, setting it equal to 0:

$$x^2 - x - 2 = 0$$

Then factor the equation:

$$(x - 2)(x + 1) = 0$$

Setting both factors equal to zero results in $x = 2$ and $x = -1$.

33. D: There will be no more coyotes when the population is 0, so set y equal to 0 and solve the quadratic equation:

$$0 = -(x-2)^2 + 1600$$

Subtract 1600 from both sides, and divide through by -1. This results in:

$$1600 = (x-2)^2.$$

Then, take the square root of both sides. This process results in the following equation:

$$\pm 40 = x - 2$$

Adding 2 to both sides results in two solutions: $x = 42$ and $x = -38$. Because the problem involves years after 2000, the only solution that makes sense is 42. Add 42 to 2000; therefore, in 2042 there will be no more coyotes.

34. B: The ball is back at the starting point when the function is equal to 800 feet. Therefore, this results in solving the equation:

$$800 = -32t^2 + 90t + 800$$

Subtract 800 off of both sides and factor the remaining terms to obtain:

$$0 = 2t(-16t + 45)$$

Setting both factors equal to 0 results in $t = 0$, which is when the ball was thrown up initially, and:

$$t = \frac{45}{16} = 2.8 \text{ seconds}$$

Therefore, it will take the ball 2.8 seconds to come back down to its starting point.

35. B: Given a rational function, the expression in the denominator can never be equal to 0. To find the domain, set the denominator equal to 0 and solve for x. This results in $2 - x = 0$, and its solution is $x = 2$. This value needs to be excluded from the set of all real numbers, and therefore the domain written in interval notation is $(-\infty, 2) \cup (2, \infty)$.

36. D: To determine whether a function is even or odd, plug $-x$ into the function. If the result is $f(x)$ the function is even, and if the result is $-f(x)$ the function is odd.

$$g(-x) = 7(-x)^3 + 5(-x) - 2 = -7x^3 - 5x - 2$$

This function is neither $f(x)$ nor $-f(x)$, so the given function is neither even nor odd.

37. A: The probability of 0.9 is closer to 1 than any of the other answers. The closer a probability is to 1, the greater the likelihood that the event will occur. The probability of 0.05 shows that it is very unlikely that an adult driver will wear their seatbelt because it is close to zero. A zero probability means that it will not occur. The probability of 0.25 is closer to zero than to one, so it shows that it is unlikely an adult will wear their seatbelt.

38. B: The goal is to first isolate the variable. The fractions can easily be cleared by multiplying the entire inequality by 5, resulting in $35 - 4x < 3$. Then, subtract 35 from both sides and divide by -4. This results in $x > 8$. Notice the inequality symbol has been flipped because both sides were divided by a negative number. The solution set, all real numbers greater than 8, is written in interval notation as $(8, \infty)$. A parenthesis shows that 8 is not included in the solution set.

39. D: This system can be solved using the method of substitution. Solving the first equation for *y* results in $y = 14 - 2x$. Plugging this into the second equation gives:

$$4x + 2(14 - 2x) = -28$$

which simplifies to $28 = -28$, an untrue statement. Therefore, this system has no solution because no *x* value will satisfy the system.

40. B: The slopes of perpendicular lines are negative reciprocals, meaning their product is equal to -1. The slope of the line given needs to be found. Its equivalent form in slope-intercept form is $y = -\frac{4}{7}x + \frac{23}{7}$, so its slope is $-\frac{4}{7}$. The negative reciprocal of this number is $\frac{7}{4}$. The only line in the options given with this same slope is:

$$y = \frac{7}{4}x - 12$$

41. C: This system can be solved using substitution. Plug the second equation in for *y* in the first equation to obtain $2x - 8x = 6$, which simplifies to $-6x = 6$. Divide both sides by 6 to get $x = -1$, which is then back-substituted into either original equation to obtain $y = -8$.

42. A: Division can be used to solve this problem. The division necessary is:

$$\frac{5.972 \times 10^{24}}{7.348 \times 10^{22}}$$

To compute this division, divide the constants first then use algebraic laws of exponents to divide the exponential expression.

This results in about 0.8127×10^2, which written in scientific notation is 8.127×10^1.

43. A: The formula for the rate of change is the same as slope: change in *y* over change in *x*. The *y*-value in this case is percentage of smokers and the *x*-value is year. The change in percentage of smokers from 2000 to 2015 was 8.1 percent. The change in x was $2000 - 2015 = -15$. Therefore:

$$8.1\%/_{-15} = -0.54\%$$

The percentage of smokers decreased 0.54 percent each year.

44. C: To solve a compound inequality, the variable must be isolated in the middle. Therefore, subtract 6 from all three parts of the inequality, and then divide all three parts by 4. This results in $-5 < x < 3$. The corresponding interval notation is (-5, 3).

45. D: Given a logarithmic function $f(x) = \log_b x$, its base is *b* and it can be written as $y = \log_b x$ with equivalent exponential equation $b^y = x$. In this problem, the base is 5, so its equivalent exponential equation is $5^y = x + 3$.

46. A: In order to solve the equation, first clear the equation from fractions by multiplying each term by 4. Therefore, the equation becomes:

$$14x + 15 + 40 = 28x + 48$$

Collecting like terms results in:

$$14x + 55 = 28x + 48$$

Subtracting 55 and $28x$ from both sides gives the equation:

$$-14x = -7$$

Dividing both sides by -14 results in the solution $x = \frac{1}{2}$. Plugging this value into the expression $6x + 7$ results in:

$$6\left(\frac{1}{2}\right) + 7 = 3 + 7 = 10$$

47. C: Set up an equation to solve for the unknown. Let t be equal to the total number of items in the grocery store. Therefore, solving for t yields:

$$\frac{1}{4}t + \frac{1}{3}t + \frac{1}{6}t + 142 = t$$

Multiplying by 12 clears the fractions, resulting in:

$$3t + 4t + 2t + 1704 = 12t$$

Collecting like terms gives:

$$9t + 1704 = 12t$$

Subtracting each side by $9t$ and the dividing by 3 gives the solution:

$$t = 568$$

Therefore, the grocery store has 568 different types of items.

48. B: The volume of a cylinder with radius r and height h is $V = \pi r^2 h$. The volume of a sphere with radius r is $V = \frac{4}{3}\pi r^3$. Therefore, the cylinder mentioned in the problem has a volume of:

$$V = \pi(6)^2 10 = 360\pi \text{ in}^3$$

The sphere has the same volume, so the equations can be set equal:

$$360\pi = \frac{4}{3}\pi r^3$$

$$\frac{360\pi}{\pi} = \frac{4}{3}r^3$$

$$r^3 = 360 \times \frac{3}{4} = 270$$

Taking the cube root of r^3 finds the radius of the sphere:

$$r = \sqrt[3]{270} = 3\sqrt[3]{10}\text{ in}$$

49. A: We can assume the volume of each container is equal. The volume of the rectangular container is:

$$v = 5 \times 9 \times 12 = 540\text{ in}^3$$

The volume of the cylindrical container is:

$$V = \pi r^2 h = \pi(2.5)^2 h = 6.25\pi h\text{ in}^3$$

Note that the radius is equal to the half of the given diameter, which is 5 inches. Setting these volumes equal results in:

$$540 = 6.25\pi h$$

Solving for h gives:

$$h = \frac{540}{6.25\pi} = \frac{86.4}{\pi} = \frac{864}{10\pi} = \frac{432}{5\pi}\text{in}$$

50. B: Foiling $(4.5 + 6.5i)(9 - 5.4i)$ results in:

$$40.5 - 24.3i + 58.5i - 35.1i^2$$

Because $i^2 = -1$, this expression is equal to:

$$40.5 + 34.2i + 35.1 = 75.6 + 35.1$$

Therefore, $a = 75.6$.

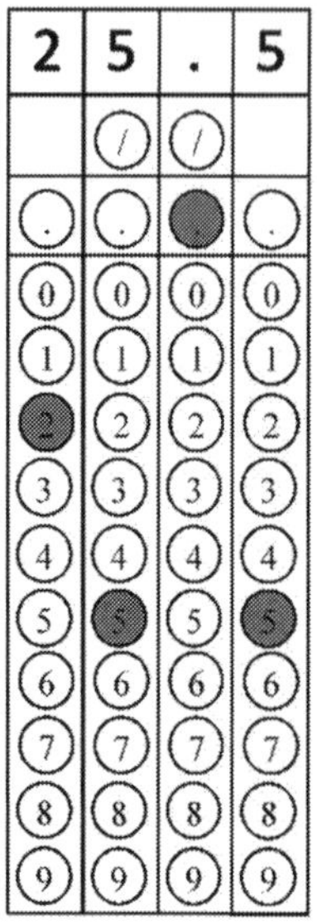

51. 25.5: In order to solve the system of equations, we multiple the second equation times 3 and add it to the first equation:

$$(3x + 9y = 18) + (3x - 9y = 45)$$

This process eliminates the variable in y and results in the following equation in x:

$$6x = 63$$

Dividing by 6 results in $x = 10.5$. Plugging this into either of the original equation results in $y = -1.5$. Therefore, the ordered pair solution to the system of equations is (10.5, -1.5). Finally, plugging this ordered pair into the given expression $3x + 4y$ results in:

$$3(10.5) + 4(-1.5) = 25.5$$

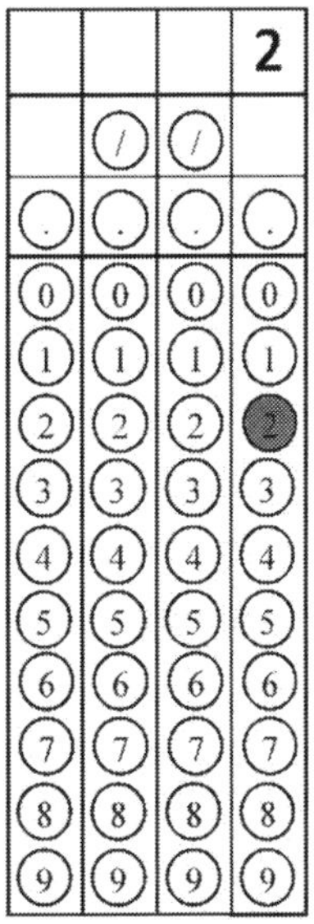

52. 2: In order for two equations to have infinitely many solutions, they must have the same slope and same y-intercept. Solving the first equation for y results in $y = -\frac{4.2}{h}x - \frac{9.1}{h}$. Solving the second equation for y results in $y = -2.1x - 4.55$. Setting these equations equal to each other and solving for h gives:

$$\frac{-4.2x}{h} - \frac{9.1}{h} = -2.1x - 4.55$$

$$\frac{-4.2x - 9.1}{h} = -2.1x - 4.55$$

$$-4.2x - 9.1 = h(-2.1x - 4.55)$$

$$h = \frac{(-4.2x - 9.1)}{(-2.1x - 4.55)}$$

$$h = 2$$

Thus, these two equations are equal when $h = 2$.

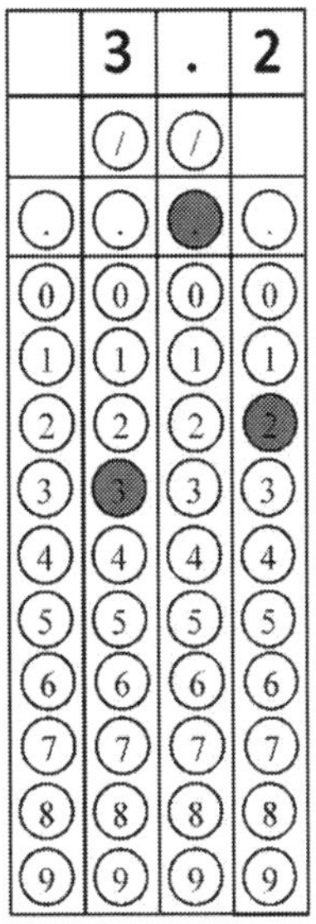

53. 3.2: The y-intercept of a line is located where the line crosses the y-axis, which is when $x = 0$. Plugging $x = 0$ in to the equation results in:

$$\frac{4}{5}(0) + \frac{5}{8}y = 2$$

Or:

$$\frac{5}{8}y = 2$$

Multiplying both sides times $\frac{8}{5}$ gives the result:

$$y = \frac{16}{5} = 3.2$$

The y-intercept is the ordered pair (0, 3.2) and its y-coordinate is 3.2.

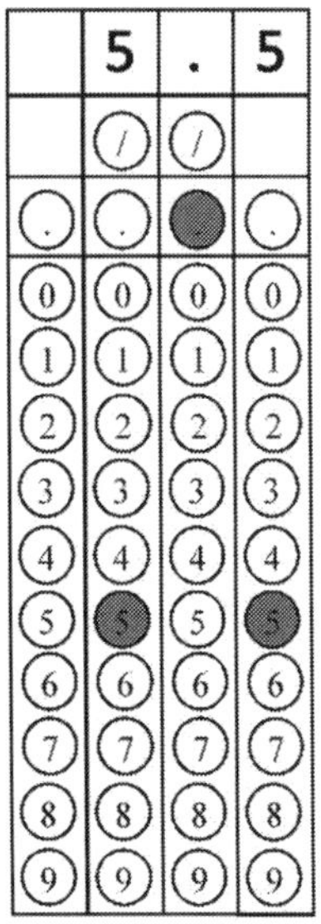

54. 5.5: To solve this system of equations, we can use the substitution method. The first equation can be rewritten as $y = 6 + x$. Plugging this in for y in the second equation results in:

$$\frac{1}{4}(3x + 9(6 + x)) = 12$$

To clear the fraction from the equation, multiply both sides times 4 to obtain:

$$3x + 9(6 + x) = 48$$

Then, distribute the 9 and collect like terms to obtain:

$$12x + 54 = 48$$

Next, subtract 54 from both sides and divide both sides by 12 to obtain $x = -\frac{1}{2} = -0.5$.

The solution for y can be obtained by plugging -0.5 in for x in the first equation:

$$-(-0.5) + y = 6$$

$$y = 5.5$$

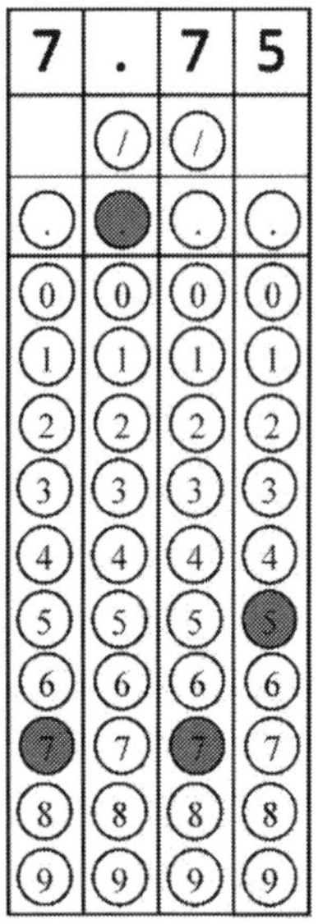

55. 7.75: Plugging $(x - c)$ into the first equation and setting equal to $4x - 22$ results in the equation $4(x - c) + 9 = 4x - 22$. Therefore, $4x - 4c + 9 = 4x - 22$. Subtracting $4x$ from both sides results in a linear equation in c: $-4c + 9 = -22$. Subtracting 9 from both sides, results in $-4c = -31$. Dividing by -4 gives the solution:

$$c = \frac{31}{4} = 7.75$$

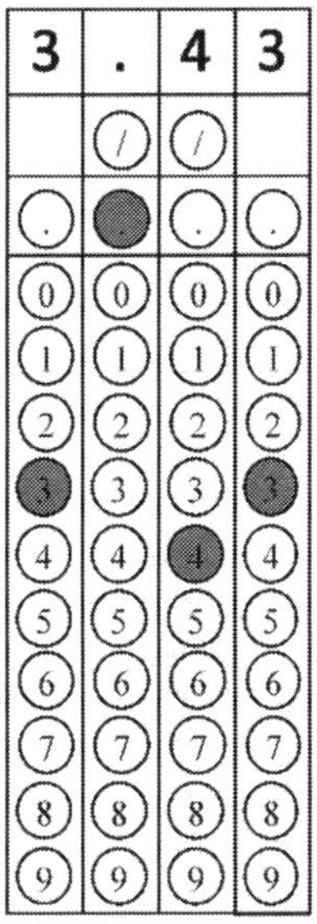

56. 3.43: First, plug 75 into the profit function to determine the profit for selling 75 pairs of shoes:

$$P(75) = 175(75) - 75^2 = 7500$$

Then, plug 60 into the profit function to determine the profit for selling 60 pairs of shoes:

$$P(60) = 175(60) - 60^2 = 6900$$

Therefore, the company will earn $7500 - 6900 = 600$ more dollars overall for charging \$15 more per pair. Because they are selling 175 pairs of shoes, the profit per pair is $\frac{600}{175}$, or approximately \$3.43 per pair.

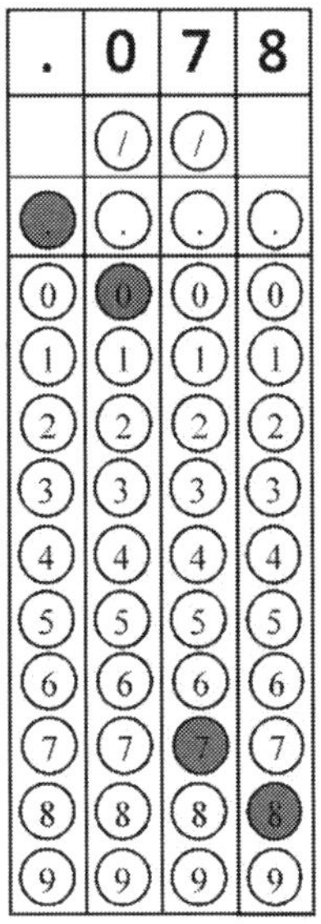

57. 0.078: Because $b^5 = -32$, $b = -2$. Plugging this into the rational expression results in:

$$\frac{-160\left(\frac{1}{(-2)^3}\right)}{((-2)^4)^2} = \frac{\frac{160}{8}}{16^2} = \frac{20}{256} \approx 0.078$$

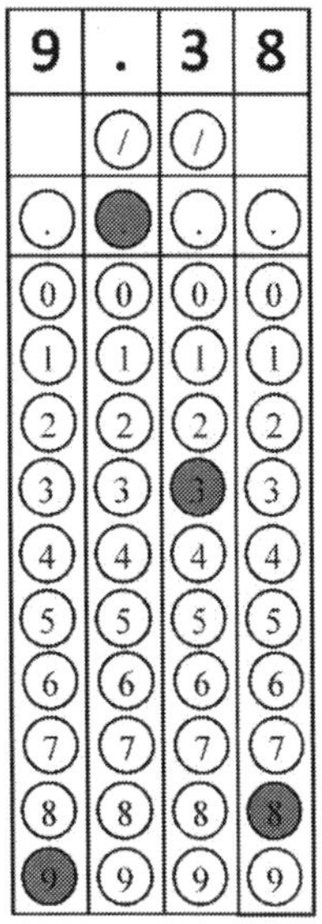

58. 9.38: This scenario represents linear versus exponential growth because the amount she spends at the coffee shop is the same each month and does not grow exponentially. The problem gives us two ordered pairs: After 8 months, she has spent \$37.50 on donuts: (8, 37.5). After 24 months, she has spent \$112.50 on donuts: (24, 112.5). The slope of the line representing this scenario is:

$$m = \frac{112.50 - 37.50}{24 - 8} = \frac{75}{16} = 4.6875$$

Therefore, she spends \$4.69 a month on donuts at the coffee shop. She spends this same amount on coffee each month, so she spends \$9.38 each month total at the coffee shop.

Practice Test #2

Reading

The following passage is taken from Chapter 6 of <u>*Sense and Sensibility*</u>*, by Jane Austen:*

The first part of their journey was performed in too melancholy a disposition to be otherwise than tedious and unpleasant. But as they drew toward the end of it, their interest in the appearance of a country which they were to inhabit overcame their dejection, and a view of Barton Valley as they entered it gave them cheerfulness. It was a pleasant fertile spot, well wooded, and rich in pasture. After winding along it for more than a mile, they reached their own house. A small green court was the whole of its demesne in front; and a neat wicket gate admitted them into it.

As a house, Barton Cottage, though small, was comfortable and compact; but as a cottage it was defective, for the building was regular, the roof was tiled, the window shutters were not painted green, nor were the walls covered with honeysuckles. A narrow passage led directly through the house into the garden behind. On each side of the entrance was a sitting room, about sixteen feet square; and beyond them were the offices and the stairs. Four bed-rooms and two garrets formed the rest of the house. It had not been built many years and was in good repair. In comparison of Norland, it was poor and small indeed!—but the tears which recollection called forth as they entered the house were soon dried away. They were cheered by the joy of the servants on their arrival, and each for the sake of the others resolved to appear happy. It was very early in September; the season was fine, and from first seeing the place under the advantage of good weather, they received an impression in its favour which was of material service in recommending it to their lasting approbation.

The situation of the house was good. High hills rose immediately behind, and at no great distance on each side; some of which were open downs, the others cultivated and woody. The village of Barton was chiefly on one of these hills, and formed a pleasant view from the cottage windows. The prospect in front was more extensive; it commanded the whole of the valley, and reached into the country beyond. The hills which surrounded the cottage terminated the valley in that direction; under another name, and in another course, it branched out again between two of the steepest of them.

With the size and furniture of the house Mrs. Dashwood was upon the whole well satisfied; for though her former style of life rendered many additions to the latter indispensable, yet to add and improve was a delight to her; and she had at this time ready money enough to supply all that was wanted of greater elegance to the apartments. "As for the house itself, to be sure," said she, "it is too small for our family, but we will make ourselves tolerably comfortable for the present, as it is too late in the year for improvements. Perhaps in the spring, if I have plenty of money, as I dare say I shall, we may think about building. These parlors are both too small for such parties of our friends as I hope to see often collected here; and I have some thoughts of throwing the passage into one of them with perhaps a part of the other, and so leave the remainder of that other for an entrance; this, with a new drawing room which may be easily added, and a bed-chamber and garret above, will make it a very snug little cottage. I could wish the stairs were handsome. But one must not expect every thing; though I suppose it would be no difficult matter to widen them. I shall see how much I am before-hand with the world in the spring, and we will plan our improvements accordingly."

In the mean time, till all these alterations could be made from the savings of an income of five hundred a-year by a woman who never saved in her life, they were wise enough to be contented with the house as it was; and each of them was busy in arranging their particular concerns, and endeavoring, by placing around them books and other possessions, to form themselves a home. Marianne's pianoforte was unpacked and properly disposed of; and Elinor's drawings were affixed to the walls of their sitting room.

In such employments as these they were interrupted soon after breakfast the next day by the entrance of their landlord, who called to welcome them to Barton, and to offer them every accommodation from his own house and garden in which theirs might at present be deficient. Sir John Middleton was a good looking man about forty. He had formerly visited at Stanhill, but it was too long for his young cousins to remember him. His countenance was thoroughly good-humoured; and his manners were as friendly as the style of his letter. Their arrival seemed to afford him real satisfaction, and their comfort to be an object of real solicitude to him. He said much of his earnest desire of their living in the most sociable terms with his family, and pressed them so cordially to dine at Barton Park every day till they were better settled at home, that, though his entreaties were carried to a point of perseverance beyond civility, they could not give offence. His kindness was not confined to words; for within an hour after he left them, a large basket full of garden stuff and fruit arrived from the park, which was followed before the end of the day by a present of game.

1. What is the point of view in this passage?
    a. Third-person omniscient
    b. Second-person
    c. First-person
    d. Third-person objective

2. Which of the following events occurred first?
    a. Sir John Middleton stopped by for a visit.
    b. The servants joyfully cheered for the family.
    c. Mrs. Dashwood discussed improvements to the cottage.
    d. Elinor hung her drawings up in the sitting room.

3. Over the course of the passage, the Dashwoods' attitude shifts. Which statement best describes that shift?
    a. From appreciation of the family's former life of privilege to disdain for the family's new landlord
    b. From confidence in the power of the family's wealth to doubt in the family's ability to survive
    c. From melancholy about leaving Norland to excitement about reaching Barton Cottage in the English countryside
    d. From cheerfulness about the family's expedition to anxiety about the upkeep of such a big home

4. Which of the following is a theme of this passage?
    a. All-conquering love
    b. Power of wealth
    c. Wisdom of experience
    d. Reality vs. expectations

5. At the start of paragraph five, the narrator says, "till all these alterations could be made from the savings of an income of five hundred a-year by a woman who never saved in her life, they were wise enough to be contented with the house as it was." What does the narrator mean?
   a. The family is going through a transition phase.
   b. Mrs. Dashwood needs to obtain meaningful employment.
   c. The family is going through a growth phase.
   d. The Dashwood children need to be concerned about the future.

6. What is the relationship between the new landlord and the Dashwoods?
   a. He is a former social acquaintance.
   b. He is one of their cousins.
   c. He is Mrs. Dashwood's father.
   d. He is a long-time friend of the family.

7. Why does the narrator describe the generosity of Sir John Middleton?
   a. To identify one of many positive traits that a landlord should possess.
   b. To explain how a landlord should conduct himself in order to be successful.
   c. To illustrate how his kindness eased the family's adaptation to their new home and circumstances.
   d. To demonstrate that he did not need to be cold and businesslike all of the time.

8. At the start of paragraph two, the narrator refers to the Dashwoods' new home as being "defective." What does the narrator mean?
   a. The tall hills surround and hide the home from neighboring structures.
   b. The building is much too poor and crowded for the family.
   c. The home's insufficiently sized parlors are too small for entertaining.
   d. The building's look and feel do not resemble that of a typical cottage.

9. Which of the following best describes the tone of the passage?
   a. Melancholy
   b. Inventive
   c. Upbeat
   d. Apprehensive

10. Toward the end of paragraph six, the narrator says, "his entreaties were carried to a point of perseverance beyond civility." What does the narrator accomplish by saying this?
   a. Signifies the Dashwoods' annoyance with the fake friendliness of Sir John Middleton
   b. Describes Sir John Middleton's slightly overbearing, overly effusive invitations
   c. Demonstrates the language from the time period in which the piece was penned
   d. Questions the genuineness of the offer for the family to stay at Barton Cottage

The following passage is taken from the Advantages of Division of Labor section in Chapter VI of *Principles of Political Economy*, *by John Stuart Mill:*

> The causes of the increased efficiency given to labor by the division of employments are some of them too familiar to require specification; but it is worthwhile to attempt a complete enumeration of them. By Adam Smith they are reduced to three: "First, the increase of dexterity in every particular workman; secondly, the saving of the time which is commonly lost in passing from one species of work to another; and, lastly, the invention of a great number of machines which facilitate and abridge labor, and enable one man to do the work of many."

(1.) Of these, the increase of dexterity of the individual workman is the most obvious and universal. It does not follow that because a thing has been done oftener it will be done better. That depends on the intelligence of the workman, and on the degree in which his mind works along with his hands. But it will be done more easily. This is as true of mental operations as of bodily. Even a child, after much practice, sums up a column of figures with a rapidity which resembles intuition. The act of speaking any language, of reading fluently, of playing music at sight, are cases as remarkable as they are familiar. Among bodily acts, dancing, gymnastic exercises, ease and brilliancy of execution on a musical instrument, are examples of the rapidity and facility acquired by repetition. In simpler manual operations the effect is, of course, still sooner produced.

(2.) The second advantage enumerated by Adam Smith as arising from the division of labor is one on which I can not help thinking that more stress is laid by him and others than it deserves. To do full justice to his opinion, I will quote his own exposition of it: "It is impossible to pass very quickly from one kind of work to another, that is carried on in a different place, and with quite different tools. A country weaver, who cultivates a small farm, must lose a good deal of time in passing from his loom to the field, and from the field to his loom. When the two trades can be carried on in the same workhouse, the loss of time is no doubt much less. It is even in this case, however, very considerable. A man commonly saunters a little in turning his hand from one sort of employment to another." I am very far from implying that these considerations are of no weight; but I think there are counter-considerations which are overlooked. If one kind of muscular or mental labor is different from another, for that very reason it is to some extent a rest from that other; and if the greatest vigor is not at once obtained in the second occupation, neither could the first have been indefinitely prolonged without some relaxation of energy. It is a matter of common experience that a change of occupation will often afford relief where complete repose would otherwise be necessary, and that a person can work many more hours without fatigue at a succession of occupations, than if confined during the whole time to one. Different occupations employ different muscles, or different energies of the mind, some of which rest and are refreshed while others work. Bodily labor itself rests from mental, and conversely. The variety itself has an invigorating effect on what, for want of a more philosophical appellation, we must term the animal spirits—so important to the efficiency of all work not mechanical, and not unimportant even to that.

(3.) The third advantage attributed by Adam Smith to the division of labor is, to a certain extent, real. Inventions tending to save labor in a particular operation are more likely to occur to any one in proportion as his thoughts are intensely directed to that occupation, and continually employed upon it.

This also can not be wholly true. "The founder of the cotton manufacture was a barber. The inventor of the power-loom was a clergyman. A farmer devised the application of the screw-propeller. A fancy-goods shopkeeper is one of the most enterprising experimentalists in agriculture. The most remarkable architectural design of our day has been furnished by a gardener. The first person who supplied London with water was a goldsmith. The first extensive maker of English roads was a blind man, bred to no trade. The father of English inland navigation was a duke, and his engineer was a millwright. The first great builder of iron bridges was a stone-mason, and the greatest railway engineer commenced his life as a colliery engineer."

(4.) The greatest advantage (next to the dexterity of the workmen) derived from the minute division of labor which takes place in modern manufacturing industry, is one not mentioned by

Adam Smith, but to which attention has been drawn by Mr. Babbage: the more economical distribution of labor by classing the work-people according to their capacity.

11. Which of the following statements would the author agree is an advantage associated with the division of labor proposed by Adam Smith?
   a. Dexterity increases as employees complete repeated tasks
   b. Repetition results in increased monotony for employees
   c. Greater interdependence forms in the production process
   d. Unemployment increases as workers are replaced by machines

12. In paragraph five, why does the author mention that the "first great builder of iron bridges was a stone-mason"?
   a. To provide an example of how a person can do anything that he or she sets his or her mind to
   b. To demonstrate that Adam Smith's third advantage associated with the division of labor is not entirely true
   c. To explain that individuals are not always employed in professions for which they have received schooling
   d. To state that it is possible for individuals to have more than one career during their working life

13. Which statement best expresses the passage's main idea?
   a. The effect of specialization of division of labor
   b. Disproving the economic principles of Adam Smith
   c. Advantages associated with the division of labor
   d. Basic principles of macroeconomics

14. Which advantage associated with the division of labor does the author say is the second most important one?
   a. Increase of dexterity in employees
   b. Savings of time by staying on one type of work
   c. Invention of machines to assist with manual labor
   d. Classification of employees by their abilities

15. In paragraph two of this passage, what does the word "bodily" mean?
   a. Tangible
   b. Animal
   c. Organic
   d. Spiritual

16. Which word best describes the author's attitude toward Adam Smith?
   a. Dismissive
   b. Respectful
   c. Adoring
   d. Questioning

17. Based on the second advantage mentioned in the passage, what would be the most efficient course of action for an employee working in a copy center?

a. The employee should focus on only a single task, such as producing black and white copies for large company orders.

b. The employee should produce copies in one room and then go to a different building hall to collate and bind documents.

c. The employee should produce black and white copies for a few hours and then change their focus to shipping out copy orders to customers.

d. The employee should be allowed unlimited flexibility to schedule their daily work tasks however they see fit.

18. Which idea is supported by this passage?

a. Repetitive tasks lead to stability in employees' work.

b. One employee should complete all of the tasks in a workflow.

c. Electronic toll collection is taking jobs from workers.

d. Milking machines help to speed up the retrieval of milk from dairy farms.

19. A doll manufacturer opens a new facility and needs employees who can cut fabric, stuff bodies, and make delicate hair and facial features. Instead of hiring employees and cross-training them on all the tasks, the manufacturer hires staff with varying degrees of skill for the various tasks. What is this is an example of?

a. Classing employees by their capacity

b. Increasing dexterity by training on repetitive tasks

c. Exclusively hiring experienced employees

d. Introducing machines to assist with unfamiliar tasks

20. In paragraph three of this passage, why does the author use the phrase "animal spirits"?

a. To provide an example of how employees feel when they work beyond fatigue

b. To explain the positive effect that variety in work has on employees

c. To demonstrate how employees feel when they do not receive proper training

d. To state how employees are aversely affected when they change tasks

Passage A

The following passage is taken from Compassion and Benevolence under The Affections in Section II Of The Philosophy of the Moral Feelings, by John Abercrombie, M.D. OXON. & EDIN:

> The exercise of the benevolent affections may be briefly treated of, under nearly the same heads as those referred to when considering the principle of Justice;—keeping in mind that they lead to greater exertion for the benefit of others, and thus often demand a greater sacrifice of self-love, than is included under the mere requirements of justice. On the other hand, benevolence is not to be exercised at the expense of Justice; as would be the case, if a man were found relieving distress by such expedients as involve the necessity of withholding the payment of just debts, or imply the neglect or infringement of some duty which he owes to another.
>
> (1.) Compassion and benevolent exertion are due toward alleviating the distresses of others. This exercise of them, in many instances, calls for a decided sacrifice of personal interest, and, in others, for considerable personal exertion. We feel our way to the proper measure of these sacrifices, by the high principle of moral duty, along with that mental exercise which places us in the situation of others, and, by a kind of reflected self-love, judges of the conduct due by us to

them in our respective circumstances.—The details of this subject would lead us into a field too extensive for our present purpose. Pecuniary aid, by those who have the means, is the most easy form in which benevolence can be gratified, and that which often requires the least, if any, sacrifice of personal comfort or self-love. The same affection maybe exercised in a degree much higher in itself, and often much more useful to others, by personal exertion and personal kindness. The former, compared with the means of the individual, may present a mere mockery of mercy; while the latter, even in the lowest walks of life, often exhibit the brightest displays of active usefulness that can adorn the human character. This high and pure benevolence not only is dispensed with willingness, when occasions present themselves; but seeks out opportunities for itself, and feels in want of its natural and healthy exercise when deprived of an object on which it may be bestowed.

(2.) Benevolence is to be exercised toward the reputation of others. This consists not only in avoiding any injury to their characters, but in exertions to protect them against the injustice of others,—to correct misrepresentations,—to check the course of slander, and to obviate the efforts of those who would poison the confidence of friends, or disturb the harmony of society.

Passage B

The following passage is taken from Part II in Section II Of Benevolence in An Enquiry Concerning the Principles of Morals, by David Hume:

Giving alms to common beggars is naturally praised; because it seems to carry relief to the distressed and indigent: but when we observe the encouragement thence arising to idleness and debauchery, we regard that species of charity rather as a weakness than a virtue.

Tyrannicide, or the assassination of usurpers and oppressive princes, was highly extolled in ancient times; because it both freed mankind from many of these monsters, and seemed to keep the others in awe, whom the sword or poignard could not reach. But history and experience having since convinced us, that this practice increases the jealousy and cruelty of princes, a Timoleon and a Brutus, though treated with indulgence on account of the prejudices of their times, are now considered as very improper models for imitation.

Liberality in princes is regarded as a mark of beneficence, but when it occurs, that the homely bread of the honest and industrious is often thereby converted into delicious cates for the idle and the prodigal, we soon retract our heedless praises. The regrets of a prince, for having lost a day, were noble and generous: but had he intended to have spent it in acts of generosity to his greedy courtiers, it was better lost than misemployed after that manner.

Luxury, or a refinement on the pleasures and conveniences of life, had not long been supposed the source of every corruption in government, and the immediate cause of faction, sedition, civil wars, and the total loss of liberty. It was, therefore, universally regarded as a vice, and was an object of declamation to all satirists, and severe moralists. Those, who prove, or attempt to prove, that such refinements rather tend to the increase of industry, civility, and arts regulate anew our MORAL as well as POLITICAL sentiments, and represent, as laudable or innocent, what had formerly been regarded as pernicious and blamable.

Upon the whole, then, it seems undeniable, THAT nothing can bestow more merit on any human creature than the sentiment of benevolence in an eminent degree; and THAT a PART, at least, of its merit arises from its tendency to promote the interests of our species, and bestow happiness on human society. We carry our view into the salutary consequences of such a character and disposition; and whatever has so benign an influence, and forwards so desirable an end, is beheld with complacency and pleasure. The social virtues are never regarded without their beneficial tendencies, nor viewed as barren and unfruitful. The happiness of mankind, the order of society, the harmony of families, the mutual support of friends, are always considered as the result of their gentle dominion over the breasts of men.

21. According to the author of passage A, which example of benevolence is the simplest to execute?
    a. Providing money to enable a student the opportunity to attend an educational workshop
    b. Holding a lemonade stand to raise funds and awareness for pediatric cancer
    c. Shopping for new toys to donate to a fundraiser that collects gifts for needy kids
    d. Volunteering to cook and serve Thanksgiving dinner for homeless people

22. In paragraph two of passage A, what kind of "mental exercise" does the author discuss?
    a. Practicing to improve an individual's recall from memory
    b. Trying to see things from another person's point of view
    c. Strengthening a person's ability to concentrate and focus
    d. Taking only a singular perspective into consideration

23. In paragraph one of passage A, what does the term "heads" mean?
    a. Discord
    b. Unlikeness
    c. Manner
    d. Opposition

24. According to passage A, which statement accurately reflects the relationship between benevolence and justice?
    a. Benevolence can be exercised at the expense of justice
    b. Acts of justice require selflessness
    c. Benevolence can be offered in lieu of payment of debts
    d. Justice can be exercised at the expense of benevolence

25. Why does the author of passage B say that giving food and money to beggars is seen as a weakness?
    a. That type of charity encourages laziness and corruption
    b. Some individuals are not truly deserving of the charity
    c. The grants cannot reach all who are affected by poverty
    d. Individuals refuse to accept the handouts due to their pride

26. In paragraph five of passage B, the author uses the phrase "their gentle dominion over the breasts of men." What has dominion over the breasts of men, according to the author?
    a. Merit
    b. The sentiment of benevolence
    c. The social virtues
    d. Complacency and pleasure

27. Which word best describes the author's attitude toward luxury in passage B?
    a. Adoring
    b. Nostalgic
    c. Objective
    d. Dismissive

28. Which statement best describes the way the two passages use point of view?
    a. Passage A is written in second-person point of view, and passage B is written in third-person objective point of view.
    b. Passage A is written in first-person point of view, and passage B is written in third-person objective point of view.
    c. Passage A is written in third-person objective point of view, and passage B is written in third-person limited omniscient point of view.
    d. Passage A is written in first-person point of view, and passage B is written in first-person point of view.

29. Which of the following statements best explains the difference in the tones of the passages?
    a. The tone of passage A is objective, and the tone of passage B is earnest.
    b. The tone of passage A is cynical, and the tone of passage B is sarcastic.
    c. The tone of passage A is excited, and the tone of passage B is ambivalent.
    d. The tone of passage A is sarcastic, and the tone of passage B is cynical.

30. The author of passage B stresses the importance of exercising benevolence for which purpose?
    a. Reducing the suffering of another person
    b. Contributing to the overall happiness of society
    c. Preventing damage to someone else's good name
    d. Attracting good karma to oneself and others

The following passage is taken from the Muscles, Tendons, and Tendon Sheaths section in Chapter XVIII of Manual of Surgery, by Alexis Thomson, F.R.C.S.ED. and Alexander Miles, F.R.C.S. ED.:

> Tendon sheaths have the same structure and function as the synovial membranes of joints and are liable to the same diseases. Apart from the tendon sheaths displayed in anatomical dissections, there is a loose peritendinous and perimuscular cellular tissue that is subject to the same pathological conditions as the tendon sheaths proper.
>
> *Tenosynovitis*: The toxic or infective agent is conveyed to the tendon sheaths through the blood-stream, as in the gouty, gonorrheal, and tuberculous varieties, or is introduced directly through a wound, as in the common pyogenic form of tenosynovitis.
>
> *Tenosynovitis crepitans:* In the simple or traumatic form of tenosynovitis, although the most prominent etiological factor is a strain or overuse of the tendon, there would appear to be some other, probably a toxic, factor in its production; otherwise the affliction would be much more common than it is: only a small proportion of those who strain or overuse their tendons become the subjects of tenosynovitis. The opposed surfaces of the tendon and its sheath are covered with fibrinous lymph, so that there is friction when they move on one another.
>
> The *clinical features* are pain on movement, tenderness on pressure over the affected tendon, and a sensation of crepitation or friction when the tendon is moved in its sheath. The crepitation may be soft like the friction of snow, or may resemble the creaking of new leather—"saddle-back creaking." There may be swelling in the long axis of the tendon, and redness and edema of the skin. If there is an effusion of fluid into the sheath, the swelling is more marked and crepitation is absent. There is little tendency to the formation of adhesions.
>
> In the upper extremity, the sheath of the long tendon of the biceps may be affected, but the condition is most common in the tendons about the wrist, particularly in the extensors of the

thumb, and it is most frequently met with in those who follow occupations which involve prolonged use or excessive straining of these tendons—for example, washerwomen or riveters. It also occurs as a result of excessive piano-playing, fencing, or rowing.

At the ankle it affects the peronei, the extensor digitorum longus, or the tibialis anterior. It is most often met with in relation to the tendo-calcaneus—*Achillo-dynia*—and results from the pressure of ill-fitting boots or from the excessive use and strain of the tendon in cycling, walking, or dancing. There is pain in raising the heel from the ground, and creaking can be felt on palpation.

The *treatment* consists in putting the affected tendon at rest, and with this object a splint may be helpful; the usual remedies for inflammation are indicated: Bier's hyperemia, lead and opium fomentations, and ichthyol and glycerin. The affliction readily subsides under treatment, but is liable to relapse on a repetition of the exciting cause.

Gouty tenosynovitis: A deposit of urate of soda beneath the endothelial covering of tendons or of that lining their sheaths is commonly met with in gouty subjects. The accumulation of urates may result in the formation of visible nodular swellings, varying in size from a pea to a cherry, attached to the tendon and moving with it. They may be merely unsightly, or they may interfere with the use of the tendon. Recurrent attacks of inflammation are prone to occur. We have removed such gouty masses with satisfactory results.

Suppurative tenosynovitis: This form usually follows upon infected wounds of the fingers—especially of the thumb or little finger—and is a frequent sequel to whitlow; it may also follow amputation of a finger. Once the infection has gained access to the sheath, it tends to spread, and may reach the palm or even the forearm, being then associated with cellulitis. In moderately acute cases the tendon and its sheath become covered with granulations, which subsequently lead to the formation of adhesions; while in more acute cases the tendon sloughs. The pus may burst into the cellular tissue outside the sheath, and the suppuration is liable to spread to neighbouring sheaths or to adjacent bones or joints—for example, those of the wrist.

The *treatment* consists in inducing hyperemia and making small incisions for the escape of pus. The site of incision is determined by the point of greatest tenderness on pressure. After the inflammation has subsided, active and passive movements are employed to prevent the formation of adhesions between the tendon and its sheath. If the tendon sloughs, the dead portion should be cut away, as its separation is extremely slow and is attended with prolonged suppuration.

31. Which statement best expresses the main idea of the passage?
    a. A discussion of the causes, symptoms, and treatments associated with various types of tenosynovitis
    b. The similarities that exist between tendon sheaths and the synovial membranes of joints
    c. An exploration of sports and professions that may be responsible for injuries to tendons
    d. Differences in how tenosynovitis displays in injuries of the wrist versus injuries of the ankle

32. According to the passage, which of the following is a treatment for tenosynovitis crepitans of the ankle?
    a. Depositing urate of soda under the endothelial tendon covering
    b. Placing the tendon at rest, possibly utilizing a splinting device
    c. Incorporating active and passive movements to prevent tendon adhesions
    d. Inducing hyperemia and making small incisions to release pus

33. Which statement is supported by the information in the third paragraph?
   a. Tenosynovitis crepitans directly results from the overuse of a tendon.
   b. Edema and redness of the skin will return a diagnosis of tenosynovitis crepitans.
   c. Since few overuse injuries result in tenosynovitis crepitans, a toxic agent may be at play.
   d. A dancer who strains a tendon will undoubtedly suffer from tenosynovitis crepitans.

34. According to the author, a patient having which of the following clinical features would be presenting with a case of Tenosynovitis crepitans?
   a. Visible nodular swellings
   b. Cellulitis infections
   c. Finger wound infections
   d. Sensation of friction

35. The authors' point of view in this passage can be described as which of the following?
   a. First-person
   b. Second-person
   c. Third-person limited omniscient
   d. Third-person objective

36. According to the passage, which of the following may be the end result of an infected wound of the finger?
   a. The formation of cellulitis
   b. Pus spreading to adjacent bones and joints of the wrist
   c. Amputation of the finger
   d. Formation of granules on tendons and sheaths

37. In paragraph four, the authors use the phrases "soft like the friction of snow" and "resemble the creaking of new leather" to describe which of the following?
   a. Pain and pressure due to an increase in movement
   b. Swelling in the long axis of the tendon
   c. Friction associated with tendon movement in the sheath
   d. Effusion of fluid in the sheath of the tendon

38. A construction worker is wearing work boots that are too tight, and he experiences pain when he lifts his heel. Based on the passage, he will most likely be diagnosed with which of the following?
   a. Tenosynovitis
   b. Suppurative tenosynovitis
   c. Gouty tenosynovitis
   d. Tenosynovitis crepitans

39. Why do the authors of this passage incorporate the words "washerwomen" and "riveters" in paragraph five?
   a. To give examples of repetitive motion jobs that may result in tenosynovitis crepitans
   b. To emphasize the importance of healthcare for blue-collar workers
   c. To provide illustrations of recurrent inflammation attacks common to gouty tenosynovitis
   d. To illustrate why ergonomics in the workplace is essential to the health of employees

40. A woman presents to her physician's office with a nodular swelling on her thumb the size of a blueberry. Based on the passage, which condition will she most likely be diagnosed with?

a. Tendo-calcaneus
b. Gouty tenosynovitis
c. Suppurative tenosynovitis
d. Tenosynovitis

The following passage is an excerpt from Rural Architecture by Lewis Falley Allen

The lover of country life who looks upon rural objects in the true spirit, and, for the first time surveys the cultivated portions of the United States, will be struck with the incongruous appearance and style of our farm houses and their contiguous buildings; and, although, on examination, he will find many, that in their interior accommodation, and perhaps relative arrangement to each other, are tolerably suited to the business and convenience of the husbandman, still, the feeling will prevail that there is an absence of method, congruity, and correct taste in the architectural structure of his buildings generally, by the American farmer.

We may, in truth, be said to have no architecture at all, as exhibited in our agricultural districts, so far as any correct system, or plan is concerned, as the better taste in building, which a few years past has introduced among us, has been chiefly confined to our cities and towns of rapid growth. Even in the comparatively few buildings in the modern style to be seen in our farming districts, from the various requirements of those buildings being partially unknown to the architect and builder, who had their planning—and upon whom, owing to their own inexperience in such matters, their employers have relied—a majority of such dwellings have turned out, if not absolute failures, certainly not what the necessities of the farmer has demanded. Consequently, save in the mere item of outward appearance—and that, not always—the farmer and cottager have gained nothing, owing to the absurdity in style or arrangement, and want of fitness to circumstances adopted for the occasion.

We have stated that our prevailing rural architecture is discordant in appearance; it may be added, that it is also uncouth, out of keeping with correct rules, and, ofttimes offensive to the eye of any lover of rural harmony. Why it is so, no matter, beyond the apology already given—that of an absence of cultivation, and thought upon the subject. It may be asked, of what consequence is it that the farmer or small property-holder should conform to given rules, or mode, in the style and arrangement of his dwelling, or out-buildings, so that they be reasonably convenient, and answer his purposes? For the same reason that he requires symmetry, excellence of form or style, in his horses, his cattle, or other farm stock, household furniture, or personal dress. It is an arrangement of artificial objects, in harmony with natural objects; a cultivation of the sympathies which every rational being should have, more or less, with true taste; that costs little or nothing in the attainment, and, when attained, is a source of gratification through life. Every human being is bound, under ordinary circumstances, to leave the world somewhat better, so far as his own acts or exertions are concerned, than he found it, in the exercise of such faculties as have been given him. Such duty, among thinking men, is conceded, so far as the moral world is concerned; and why not in the artificial? So far as the influence for good goes, in all practical use, from the building of a temple, to the knocking together of a pig-stye—a labor of years, or the work of a day—the exercise of a correct taste is important, in a degree.

41. What is the main idea of this passage?
   a. Farmers need protection from the government in keeping their farmhouses, and they need more grant money to rebuild better farms.
   b. Urban architecture is better than rural architecture because more funds are being poured into the building of cities.
   c. Rural architecture in America is inconsistent in style and construct and is thus offensive to the farmer's need of symmetry and order.
   d. When you compare the style of European farms to American farms, one can see that American farms are superior in their construction.

42. What is the author's tone in the passage?
   a. Passionate
   b. Critical
   c. Neutral
   d. Joyous

43. What is the meaning of the word *incongruous* at the beginning of the third paragraph?
   a. Out of place
   b. Perfectly symmetrical
   c. Having abundance
   d. Having the characteristic of beauty

44. What does the author say about when human beings leave the world?
   a. They should leave it better than when they found it.
   b. They should look back on the architecture and be proud of it.
   c. They are responsible for the debts they leave behind.
   d. Their family will look to the work they have left behind.

45. What does the author imply about America's rural architecture?
   a. It is brilliant and cultivated
   b. It has great potential
   c. It is humble and practical
   d. It is poor and uncivilized

46. What is the meaning of the word *discordant* at the beginning of the third paragraph?
   a. Pleasantness
   b. Harmonizing
   c. Conflicting
   d. Bold

The following passage is an excerpt from Quilts: Their Story and How to Make Them by Marie D. Webster

> Quilts, as known and used to-day, may be divided into two general classes, washable and non-washable, depending upon the materials of which they are made. The methods for constructing each class are the same, and are so very simple that it seems hardly necessary to explain them.
>
> The name quilt implies two or more fabrics held together with many stitches. Webster defines a quilt as "Anything that is quilted, especially as a quilted bedcover or a skirt worn by women; any cover or garment made by putting wool, cotton, etc., between two cloths and stitching them together." The verb, to quilt, he defines as "To stitch or to sew together at frequent intervals in

order to confine in place the several layers of cloth and wadding of which a garment, comforter, etc., may be made. To stitch or sew in lines or patterns."

The "Encyclopædia Britannica" is a little more explicit and also gives the derivation of the name, quilt, as follows: "Probably a coverlet for a bed consisting of a mass of feathers, down, wool, or other soft substances, surrounded by an outer covering of linen, cloth, or other material." In its earlier days the "quilt" was often made thick and sewed as a form of mattress. The term was also given to a stitched, wadded lining for body armour. "The word came into English from old French *cuilte*. This is derived from Latin *culcitra*, a stuffed mattress or cushion. From the form *culcitra* came old French *cotra*, or *coutre* whence *coutre pointe*, this was corrupted into counterpoint, which in turn was changed to counterpane. The word 'pane' is also from the Latin *pannus*, a piece of cloth. Thus 'counterpane,' a coverlet for a bed, and 'quilt' are by origin the same word."

Broadly speaking, from these definitions, any article made up with an interlining may be called a quilt. However, usage has restricted the meaning of the word until now it is applied to a single form of bed covering. In the United States the distinction has been carried even farther and a quilt is understood to be a light weight, closely stitched bedcover. When made thicker, and consequently warmer, it is called a "comfort."

47. What is the main purpose of this passage?
    a. To persuade the audience to buy a quilt.
    b. To explain the definition and etymology of quilts.
    c. To praise quilters for their hard work and skill.
    d. To tell a story about the history of quilts.

48. What kind of evidence does the author use in the passage?
    a. Illustrations
    b. Statistics
    c. Paraphrase
    d. Direct quotes

49. What does the author say when comparing the two dictionary definitions of the word *quilt*?
    a. The author says that Webster goes off track with its definition and is incorrect.
    b. The author says that Webster gives a more specific definition than the Encyclopedia Britannica.
    c. The author says that the Encyclopedia Britannica gives a more specific definition than Webster.
    d. The author says that the Encyclopedia Britannica goes off track with its definition and is incorrect.

50. What does the author say about comforters, or the "comfort"?
    a. Comforts refer to the cotton between the two blankets being stitched together.
    b. Comforts are when family comes to you when you're in a time of need.
    c. Comforts are blankets that are stitched together that are thicker than quilts.
    d. Comforts are a type of eggnog that people make during the holidays.

51. According to the passage, which of the following words has the same origin as *quilt*?
    a. Counsel
    b. Counterpane
    c. Question
    d. Query

52. What type of passage is this?
  a. Informative
  b. Persuasive
  c. Narrative
  d. Cause and Effect

Writing and Language

Questions 1–9 are based on the following passage:

When Nathaniel Hawthorne wrote *The Scarlet Letter* in 1850, he became the first American author to contribute a mature tragic view of life (1) in the literature of the young United States. He developed a tragic vision that is fundamentally dark: man's inevitable moral downfall is determined by the workings of (2) its own heart. Yet Hawthorne's bleak vision also embraces the drama of the human heart, through which he depicts the heart as a redemptive agent, subtly affirming (3) humanities inherent goodness. He likens the human heart to a cavern. To a visitor, the mouth of the cavern is bright with sunlight and flowers, (4) but a few feet in, the light dims and warmth turns to chill; the visitor stumbles, first in confusion, then in terror. (5) Further back, a small gleam of light appears and the visitor hurries toward it, to find a scene much like that at the entrance of the cavern, only perfect. This, to Hawthorne, is the depth of (6) human nature; the beauty that lies beyond fear and hopelessness.

The novel relates the suffering of three people—Hester Prynne, Arthur Dimmesdale, and Roger Chillingworth—each possessing a proud heart that leads them to a choice between a natural morality (represented by the passion that derives from human love) (7) and unnatural morality (symbolized by the code of punishment the Puritan community adheres to). Studying the heart in the context of seventeenth-century Puritanism, Hawthorne focuses on the contradictions between the theological and intellectual definitions of sin, according to the Puritans, and the (8) psychological affects of applying them to the reality of their actions and lives. Within the (9) rigid moral setting of the New England Puritan community, he concentrates upon the sin of pride, fostered in the dim, shadowy hollows of the heart and soul.

1. The best replacement for the underlined portion would be:
  a. NO CHANGE
  b. to
  c. from
  d. with

2. The best replacement for the underlined portion would be:
  a. NO CHANGE
  b. her own heart
  c. his own heart
  d. people's own heart

3. The best replacement for the underlined portion would be:
   a. NO CHANGE
   b. humanities'
   c. humanitys
   d. humanity's

4. The best replacement for the underlined portion would be:
   a. NO CHANGE
   but a few feet in, the light dims and warmth turns to chill; the visitor stumbles first in confusion, then in terror.
   b. but a few feet in the light, dims and warmth turns to chill, the visitor stumbles, first in confusion, then in terror
   c. but a few feet in the light dims and warmth turns to chill, the visitor stumbles first in confusion then in terror
   d. but a few feet in the light dims and warmth turns to chill: the visitor stumbles first in confusion, then in terror

5. The best replacement for the underlined portion would be:
   a. NO CHANGE
   b. in the back
   c. to the back
   d. farther in

6. The best replacement for the underlined portion would be:
   a. NO CHANGE
   b. human nature. The beauty that lies beyond fear and hopelessness.
   c. human nature: the beauty that lies beyond fear and hopelessness.
   d. human nature (the beauty that lies beyond fear and hopelessness).

7. The best replacement for the underlined portion would be:
   a. NO CHANGE
   b. between a natural morality (represented by the passion that derives from human love) and an unnatural morality (symbolized by the code of punishment the Puritan community adheres to).
   c. between a natural morality represented by the passion that derives from human love and unnatural morality symbolized by the code of punishment the Puritan community adheres to.
   d. between a natural morality—represented by the passion that derives from human love—and unnatural morality, symbolized by the code of punishment the Puritan community adheres to.

8. The best replacement for the underlined portion would be:
   a. NO CHANGE
   b. psychological damage
   c. psychological changes
   d. psychological effects

9. Which choice most accurately conveys the meaning of the underlined word?
   a. Frightening
   b. Strict
   c. Harsh
   d. Dangerous

Questions 10–18 are based on the following passage:

The vast (10) industrialization of Europe which took place between 1760 and 1840 was perhaps the most significant watershed period in the history of Europe. It was a time of astronomical growth and progress for Europeans as a whole, yet it was simultaneously a wretched and dehumanizing period for the majority of European individuals. Many problems—economic, social, and political—were either created or magnified by the Industrial Revolution: (11) problems which threatened European society not with annihilation but with massive change in the form of a social revolution that would shift the distribution of political might and change class structure.

The Industrial Revolution touched and altered almost every aspect of the economic and political life of Europe prior to 1760, which, in turn, (12) changed the existing social order. (13) With roots digging deeply into the past as the thirteenth century, when capitalism and commerce began to develop, industrialization slowly became (14) inevitable. It was aided by a gradual expansion of the market, a demand for more goods by an increasing number of consumers, and the step-by-step freedom of private enterprise from government control. (15) Industrialization began in earnest in the textile industry of Great Britain, and, due to progress in the field of technological innovations, caused a huge upswing in the amount of money necessary to establish a factory. Industry became the new source of wealth, (22) which had formerly been land, but the power remained concentrated within a small group of rich men—the capitalists. The existence of this wealthy class directly contrasted with that of the impoverished working class (people who had once worked the land, until they were forced by industrialization to undertake factory labor in order to survive). The juxtaposition of a small (23) number of people, who held great wealth, with a population for whom extreme and widespread poverty was inescapable, was one of the greatest problems the Industrial Revolution created.

The plight of the industrial family was a shameful aspect of the modernization of Europe. Forced to migrate to areas rich in coal, (16) where filthy, violence-ridden cities have sprung up, the people of agrarian communities encountered suffering, despondency, and poverty. Wages were perpetually too low, and employment was always uncertain. Housing was damp, dirty, cramped, and poorly ventilated, leading to sickness and the spread of disease. Death by starvation was not uncommon, as food was scarce and often unfit to eat.

10. The best replacement for the underlined portion would be:
   a. NO CHANGE
   b. industrialization of Europe, which took place between 1760 and 1840 was perhaps the most significant watershed
   c. industrialization of Europe, which took place between 1760 and 1840, was perhaps the most significant watershed
   d. industrialization of Europe which took place, between 1760 and 1840, was perhaps the most significant watershed

11. The best replacement for the underlined portion would be:
   a. NO CHANGE
   b. problems threatened European society
   c. problems threatening to European society
   d. problems that threatened European society

12. In the context of the essay as a whole, which of the following words is the most accurate substitution for the underlined portion?
  a. transfigured
  b. disrupted
  c. destroyed
  d. transformed

13. The best replacement for the underlined portion would be:
  a. NO CHANGE
  b. With roots, digging deeply into the past as the thirteenth century
  c. With roots digging as deeply into the past as the thirteenth century
  d. With roots that were digging deeply into the past as the thirteenth century

14. Is the underlined word the most appropriate word choice?
  a. Yes, "inevitable" is most appropriate.
  b. No, "indelible" is more appropriate.
  c. No, "inadvertent" is more appropriate.
  d. No, "inadvisable" is more appropriate.

15. A transition sentence is needed to improve the flow from the sentence with number 21 to the following sentence. Which of the following sentences best accomplishes the transition?
  a. The demand for large amounts of capital was met by a small group of wealthy investors.
  b. Therefore, the need of large capitalists was predominant.
  c. Capitalists, with their deep pockets, came to the rescue.
  d. The economic theory of supply and demand was born.

16. The underlined sentence is awkward. Which choice below is the clearest rewrite?
  a. Land, formerly the source of wealth, was overshadowed by industry, dominated by rich capitalists.
  b. Capitalists invested in industry, which replaced land as the primary source of wealth. As a result, power was concentrated in the hands of the capitalists.
  c. By investing their capital in industry, the capitalists changed the source of wealth from land to industry and retained most of the power.
  d. Formerly the major source of wealth, land was replaced by industry, and the capitalists held all the power.

17. The best replacement for the underlined portion would be:
  a. NO CHANGE
  b. amount
  c. group
  d. class

18. The best replacement for the underlined portion would be:
  a. NO CHANGE
  b. where filthy, violence-ridden, cities have sprung up
  c. where filthy, violence-ridden cities had sprung up
  d. where filthy, violence ridden cities have sprung up

Questions 19–27 are based on the following passage:

The escalating rate of obesity in the United States is a major health issue facing all healthcare providers and (19) <u>faculties</u>. Today, at Carter Cullen Medical Center, a community hospital in (20) <u>Wayland, Massachusetts</u>, a simple perioperative step—highlighting body mass index (BMI, determined by dividing weight by height) on surgery schedules and preoperative checklists—is saving lives. In 2014, events involving two patients with high (21) <u>BMI's</u> triggered eighteen months of focused teamwork and communication that resulted in the establishment of a groundbreaking early notification protocol. As a consequence, the healthcare professionals at Carter Cullen Medical Center are better equipped to care for patients who are morbidly obese.

Obesity carries many risks, one of which is sleep apnea. (22) <u>Sleep apnea, in turn, causes problems with intubation during surgery and disrupts the administration of anesthetics</u> or oxygen. Additional adipose tissue in the neck area prevents patients from hyperextending their necks properly. During intubation, the anesthesiologist cannot visualize the vocal cords (23) <u>if the patients neck</u> is not hyperextended, and a fiber-optic intubation device must be inserted. But such a device was not regularly available in Carter Cullen's operating rooms; (24) <u>the equipment was stocked only if difficulty was expected</u>. In the first event, an obese patient needed to be intubated, there was unanticipated difficulty with the intubation, and the patient died.

Another life-threatening risk of obesity is rhabdomyolysis. In the second event, an obese patient recovering from prolonged total joint replacement surgery was resting in one position for an (25) <u>unreasonable</u> amount of time. The patient's weight caused muscle fibers (26) <u>to break down, and release a pigment—harmful to the kidneys—</u>into (27) <u>their</u> circulation. Renal failure and death followed.

19. Which, if any, of the following choices is the most accurate choice to replace the underlined word?
    a. NO CHANGE
    b. facilities
    c. cardiologists
    d. the American Obesity Association

20. Which, if any, of the following choices is the most accurate choice to replace the underlined words?
    a. NO CHANGE
    b. Wayland Massachusetts
    c. Wayland, Massachusets
    d. a small Massachusetts town

21. Which, if any, of the following choices is the most accurate choice to replace the underlined word?
    a. NO CHANGE
    b. BMIs
    c. body mass indexes
    d. body mass indices

22. Which, if any, of the following choices is the most accurate choice to replace the underlined words?
    a. NO CHANGE
    b. Sleep apnea, in turn, caused problems with intubation during surgery and disrupted the administration of anesthetics
    c. Sleep apnea, in turn, can cause problems with intubation during surgery that disrupt the administration of anesthetics
    d. Sleep apnea, in turn, causes problems with intubation and the administration of anesthetics

23. Which, if any, of the following choices is the most accurate choice to replace the underlined words?
    a. NO CHANGE
    b. if the patient neck
    c. if the patient's neck
    d. if the patients' neck

24. Which, if any, of the following choices is the most accurate choice to replace the underlined words?
    a. NO CHANGE
    b. the equipment was stocked if difficulty only was expected
    c. the equipment only was stocked if difficulty was expected
    d. only the equipment was stocked if difficulty was expected

25. Within the context of this paragraph, which word makes the most sense?
    a. NO CHANGE
    b. outrageous
    c. unacceptable
    d. extended

26. Which choice clarifies the underlined passage and is grammatically correct?
    a. to break down, and release a pigment harmful to the kidneys
    b. to break down, and release a pigment—harmful to the kidneys
    c. to break down and release a pigment harmful to the kidneys
    d. to break down—and release a pigment—harmful to the kidneys—

27. Which of the following words is a grammatically correct replacement for the underlined portion?
    a. NO CHANGE
    b. They're
    c. There
    d. Its

Questions 28–36 are based on the following passage:

> Cerebrovascular accident, commonly known as stroke, is the fifth leading cause of death in the United States, and it (28) affects approximately 795,000 Americans each year. (29) The risk of stroke varies by demographics and lifestyle factors such as age (fifty years or older), gender (female), race (African American, American Indian, and Alaskan native), geography (southeastern states), disease, and lifestyle components (including, but not limited to: exercise, eating habits, use of alcohol and/or illicit drugs, cigarette smoking, stress, quality of life, and perceived happiness).
>
> (30) What exactly is a stroke? It is a sudden interruption in cerebral (brain) function that (31) lasts more than twenty-four hours. Stroke can result in death, and is caused by an acute disruption of

blood flow coupled with insufficient oxygen reaching the brain. A stroke can be large and catastrophic, ending in death; mild, with the person who had the stroke eventually recovering most normal functions; or anywhere between these two (32) extremes.

The two main types of stroke are ischemic and hemorrhagic. Ischemic stroke occurs when a blood clot blocks an artery in the central nervous system (CNS). (33) The blockage disrupts the flow of oxygen-rich blood to CNS tissue and to the brain to the extent that the supply of oxygen and glucose is insufficient to support ongoing metabolism. Thrombotic stroke and embolic stroke are two subtypes of ischemic stroke. In thrombotic stroke, a blood clot or atherosclerotic plaque develops (34) locally to create blockage. (35) In embolic stroke, a blood clot or atherosclerotic plaque develops elsewhere, such as in the heart, breaks apart, and travels in the bloodstream without blocking blood flow until it reaches a central nervous system artery, then it travels through the artery up to the brain, lodging there.

(36) Headache, nausea, vomiting, numbness, and loss of consciousness are all warning signs of a possible stroke these symptoms require immediate medical attention to prevent damage to—or complete loss of—brain cells. The three most common symptoms of stroke are varying degrees of facial drooping, unilateral limb weakness, and speech difficulty. Strokes can also cause depression and a loss of the ability to control emotions.

28. Which choice, if any, would be an acceptable substitute for the underlined portion?
   a. destroys
   b. effects
   c. kills
   d. afflicts

29. Carlos is a 27-year-old Hispanic male who lives in New York, and he maintains a healthy lifestyle by eating wholesome foods, exercising regularly, and getting an average of eight hours of sleep per night. He enjoys an occasional beer or glass of wine on weekends, but he does not drink to excess and has never used illicit drugs. He does not smoke, he practices yoga and meditation to control stress, and he is generally satisfied with where he is in his life today. According to the underlined passage, Carlos:
   a. will never have a stroke if he continues living as he does now
   b. will never have a stroke because he is not a "match" in any of the categories
   c. will have a stroke if he stops exercising, gains weight, and starts smoking cigarettes
   d. might be one of the Americans who has a stroke

30. If this sentence were added at the end of Paragraph Number 2, what would the effect be? "The prevalence of pseudobulbar affect among stroke survivors is estimated to be between eleven and fifty-two percent."
   a. It would be confusing because the information is irrelevant to the essay.
   b. It would be helpful because it adds information that supports the topic of the paragraph.
   c. It would be confusing because pseudobulbar affect has not been introduced previously.
   d. It would be helpful because it cites a relatively positive stroke survival statistic.

31. What is the most logical ordering of the three items in the underlined passage?
  a. NO CHANGE
  b. is caused by an acute disruption of blood flow, lasts more than twenty-four hours, and can result in death
  c. can result in death, lasts more than twenty-four hours, and is caused by an acute disruption of blood flow
  d. can result in death, is caused by an acute disruption of blood flow, lasts more than twenty-four hours

32. In the context of the sentence, which word is the closest synonym to the underlined word?
  a. Spectrums
  b. States
  c. Poles
  d. Positions

33. Where is the best place to insert a comma in the underlined passage?
  a. The punctuation is correct as is.
  b. The blockage disrupts the flow of oxygen-rich blood to CNS tissue and to the brain, to the extent that the supply of oxygen and glucose is insufficient to support ongoing metabolism.
  c. The blockage disrupts the flow of oxygen-rich blood to CNS tissue, and to the brain to the extent that the supply of oxygen and glucose is insufficient to support ongoing metabolism.
  d. The blockage disrupts the flow of oxygen-rich blood, to CNS tissue and to the brain to the extent that the supply of oxygen and glucose is insufficient to support ongoing metabolism.

34. In the context of this paragraph, what does the word "locally" word mean?
  a. In the brain
  b. In the CNS
  c. Anywhere in the arteries where there are clots and plaque
  d. In the fat cells

35. There is a semicolon missing from this run-on sentence. What is the correct position for it?
  a. In embolic stroke, a blood clot or atherosclerotic plaque develops elsewhere, such as in the heart; breaks apart, and travels in the bloodstream without blocking blood flow until it reaches a central nervous system artery, then it travels through the artery up to the brain, lodging there.
  b. In embolic stroke, a blood clot or atherosclerotic plaque develops elsewhere, such as in the heart, breaks apart, and travels in the bloodstream without blocking blood flow until it reaches a central nervous system artery; then it travels through the artery up to the brain, lodging there.
  c. In embolic stroke, a blood clot or atherosclerotic plaque develops elsewhere; such as in the heart, breaks apart, and travels in the bloodstream without blocking blood flow until it reaches a central nervous system artery, then it travels through the artery up to the brain, lodging there.
  d. In embolic stroke, a blood clot or atherosclerotic plaque develops elsewhere, such as in the heart, breaks apart; and travels in the bloodstream without blocking blood flow until it reaches a central nervous system artery, then it travels through the artery up to the brain, lodging there.

36. Which punctuation mark is the correct one to place between "stroke" and "these"?
  a. Slash (/)
  b. Dash (—)
  c. Semicolon (;)
  d. Period (.)

Questions 37–45 are based on the following passage:

The concept of three-ness, (37) embodied by triplets, triads, and trios, is crucial to Indo-European societies as a way of creating order and stability within the social structure. (38) The significance of the notion of three-ness is manifested in the literature of these cultures, which abounds with numerous events, occurring in patterns of three, that radically affect the lives of the main characters and (39) those of there families, friends, and foes. In no body of literature, perhaps, is this more (40) distinctly and consistently seen than in that of the early Icelandic (41) inhabitants. (42) The sagas produced by the Icelanders are surfeited with images of three, minor and major; it is in the later, and most widely acclaimed sagas (*The Saga of Gisli*, *Laxdaela Saga*, and *Njal's Saga*), that three-ness is applied most skillfully.

Saga authors exercised authorial control by using three-ness to create a balanced, rhythmic, organically unified fiction grounded in historical fact. (43) There are two levels of significance in how they applied the concept of three. The first level consists of instances in which the author mentions the number three (or multiples of three) that are minor in import. For example, three is widely used to name the number of sons the saga-man has, or the number of days he spends visiting a friend. Small multiples of three, such as fifteen and eighteen, mark the number of warriors fighting on either side in a skirmish, while greater multiples, such as sixty, 120, and three hundred are predominately associated with wealth or livestock. (44) These recurring instances of three-ness serve a dual purpose. (45) Not only do they create a regular, periodic cadence—a steady underlying tempo—that links all of the sections of the saga together, but they also provide, by means of the tripod-like sense of proportion, an equilibrium associated with the number three: a satisfying feeling of balance.

37. The definition of the underlined word is "to show in concrete form," and it is not entirely appropriate in this context. Which choice is most accurate?
   a. Marked
   b. Earmarked
   c. Included
   d. Represented

38. Which word or phrase should be eliminated to make the sentence less wordy?
   a. The significance of
   b. the notion of
   c. of these cultures
   d. numerous

39. If there is an error in the underlined sentence, what is the error?
   a. NO ERROR
   b. "Abounds" should be "abound."
   c. "There" should be "their."
   d. There should not be a comma before "and foes."

40. Choose the pair of words that is the closest substitute for "distinctly and consistently."
   a. Discerningly and constantly
   b. Clearly and uniformly
   c. Definitely and evenly
   d. Easily and transparently

41. What is the closest synonym to the underlined word?
   a. Dwellers
   b. Authors
   c. Cultures
   d. Persons

42. This sentence originally appeared at the end of the first paragraph, but it was later deleted: "The sagas are so noticeably saturated with trinities, threesomes, and treble happenings that the modern reader almost comes away from reading them with the numeral three indelibly stamped on her consciousness." What is the best reason for its deletion?
   a. It says the modern reader is female, which excludes the ones who are male.
   b. "Trinities, threesomes, and treble happenings" echoes the first sentence's "triplets, triads, and trios" and seems a bit forced, especially "treble happenings."
   c. What happens to the modern reader is irrelevant.
   d. The writer cannot possibly know the impact the repetition of threes has on anyone but his or her own self.

43. Which sentence would be the clearest, if inserted in this paragraph?
   a. There are two levels of significance to how they applied the concept of three.
   b. How they applied the concept of three has two levels.
   c. They applied the concept of three on two significant levels.
   d. There are two levels of significance to the application of three.

44. The underlined passage begins a new topic. This passage and the rest of the paragraph would work best as a separate paragraph. Where should the new paragraph be located?
   a. Immediately after the paragraph it is currently in
   b. Immediately after the first paragraph
   c. Immediately before the last paragraph
   d. Immediately after the last paragraph

45. Which is the clearest way of breaking this sentence into two sentences?
   a. They create a regular, periodic cadence—a steady underlying tempo—that links all of the sections of the saga together. They also provide, by means of the tripod-like sense of proportion, an equilibrium associated with the number three: a satisfying feeling of balance.
   b. Not only do they create a regular, periodic cadence that links all of the sections of the saga together. They also provide, by means of the tripod-like sense of proportion, an equilibrium associated with the number three.
   c. Not only do they create a regular, periodic cadence—a steady underlying tempo—that links all of the sections of the saga together. But they also provide, by means of the tripod-like sense of proportion, an equilibrium associated with the number three: a satisfying feeling of balance.
   d. A regular, periodic cadence—a steady underlying tempo—that links all of the sections of the saga together is created. Also, an equilibrium associated with the number three, a satisfying feeling of balance, is provided by means of the tripod-like sense of proportion.

Math

No Calculator Questions

1. What is the volume of the cylinder below?

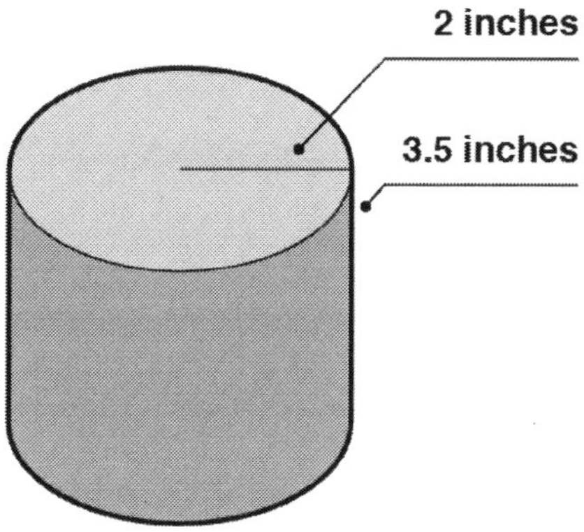

a. 18.84 in^3
b. 45.00 in^3
c. 70.43 in^3
d. 43.96 in^3

2. What is the domain of the logarithmic function $f(x) = \log_2(x-2)$?
a. 2
b. $(-\infty, \infty)$
c. $(0, \infty)$
d. $(2, \infty)$

3. What is an equivalent logarithmic form of the following exponential function?

$$f(x) = 10^{x+3}$$

a. $x = 3 + \log y$
b. $-3 + \log y = x$
c. $\log(x-3) = y$
d. $\log_3 x = \mathrm{y}$

4. What type of units are used to describe surface area?
a. Square
b. Cubic
c. Single
d. Quartic

5. Given the function $f(x) = 4x - 2$, what is the correct form of the simplified difference quotient:

$$\frac{f(x+h) - f(x)}{h}$$

a. $4x - 1$
b. $4x$
c. 4
d. $4x + h$

6. Which expression is equivalent to $\sqrt[4]{x^6} - \frac{x}{x^3} + x - 2$?

a. $x^{\frac{3}{2}} - x^2 + x - 2$

b. $x^{\frac{2}{3}} - x^{-2} + x - 2$

c. $x^{\frac{3}{2}} - \frac{1}{x^2} + x - 2$

d. $x^{\frac{2}{3}} - \frac{1}{x^2} + x - 2$

7. What is an equivalent form of the rational expression $\sqrt{200x^6y^7z^2}$? Assume all variables represent positive real numbers.

a. $20x^3y^3z\sqrt{y}$

b. $10x^3y^3z\sqrt{2y}$

c. $10xyz\sqrt{2xyz}$

d. $10x^2y^3z\sqrt{2z}$

8. What is the equation of the line that passes through the two points (-3, 7) and (-1, -5)?

a. $y = 6x + 11$

b. $y = 6x$

c. $y = -6x - 11$

d. $y = -6x$

9. Which of the following is the equation of a vertical line that runs through the point (1, 4)?

a. $x = 1$

b. $y = 1$

c. $x = 4$

d. $y = 4$

10. What is the correct result when rationalizing the denominator of $\frac{49}{x-\sqrt{7}}$?

a. $\frac{49}{x^2-7}$

b. $\frac{49x+49\sqrt{7}}{x^2-7}$

c. $\frac{x}{x^2-49}$

d. $\frac{49}{x}$

11. Given the function $h(x) = -5x - 7$, what is the value of $h(-5x)$?

a. $-25x - 7$

b. $25x - 7$

c. $25x^2 - 7$

d. 18

12. Consider the function $g(x) = ax^2 + 4x + 9$. It is true that $g(2) = -7$ and a is a constant. What is the value of a?

a. -6
b. 6
c. -7
d. 12

13. Which is equivalent to the following expression?

$$\frac{4x-2}{x+5} - \frac{2x-1}{x-5}$$

a. $\frac{6x^2-12x+5}{x^2-25}$
b. $\frac{2x-1}{x^2-25}$
c. $\frac{2x-1}{x+5}$
d. $\frac{2x^2-31x+15}{x^2-25}$

14. If $M = 5x^2 - 4x + 2$ and $N = -2x^2 - 6x + 1$, which of the following is $3M - \frac{3}{4}N$?

a. $\frac{33}{2}x^2 - \frac{15}{2}x + \frac{21}{4}$
b. $\frac{27}{2}x^2 - \frac{21}{2}x + \frac{27}{4}$
c. $\frac{33}{2}x^2 - 18x + 7$
d. $\frac{27}{2}x^2 + \frac{21}{x} + \frac{27}{4}$

15. Which of the expressions is equal to B for the equation below?

$$B + \frac{4}{9x+2} = \frac{81x^2}{9x+2}$$

a. $9x + 2$
b. $81x^2$
c. $9x - 2$
d. $81x^2 - 4$

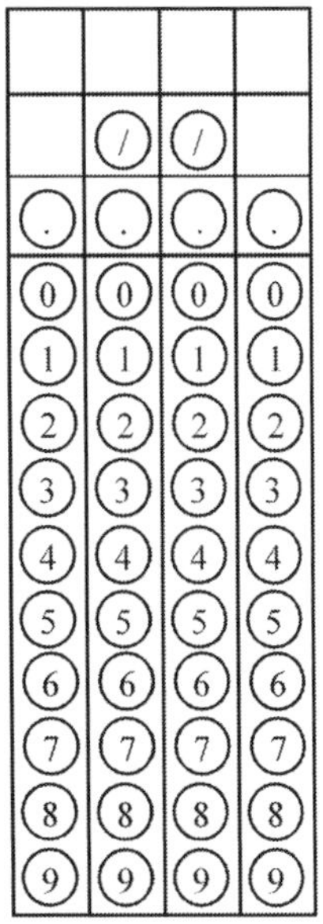

16. What is the area of the shaded region?

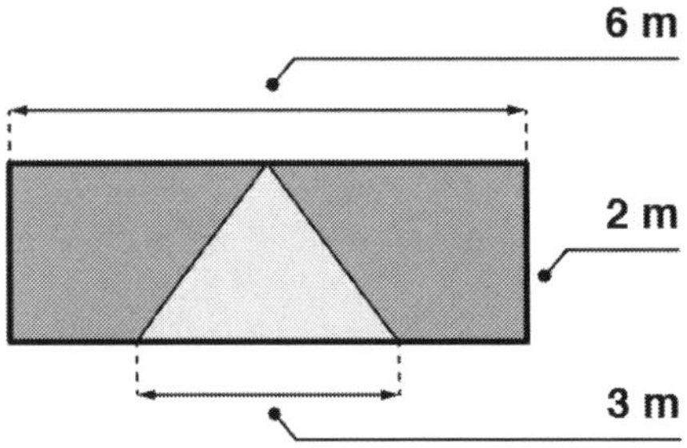

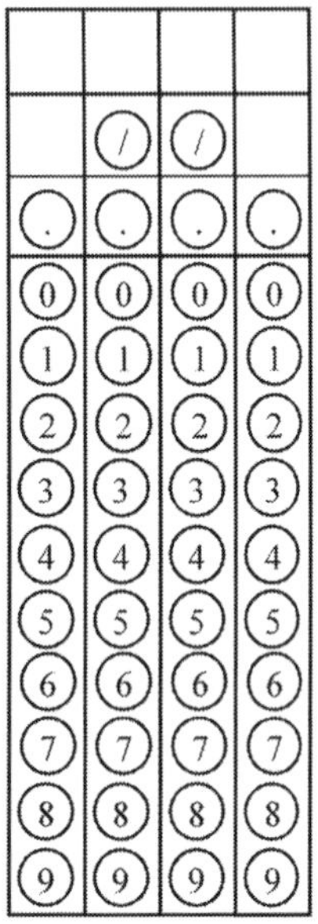

17. How many possible positive zeros does the polynomial function: $f(x) = x^4 - 3x^3 + 2x + x - 3$ have?

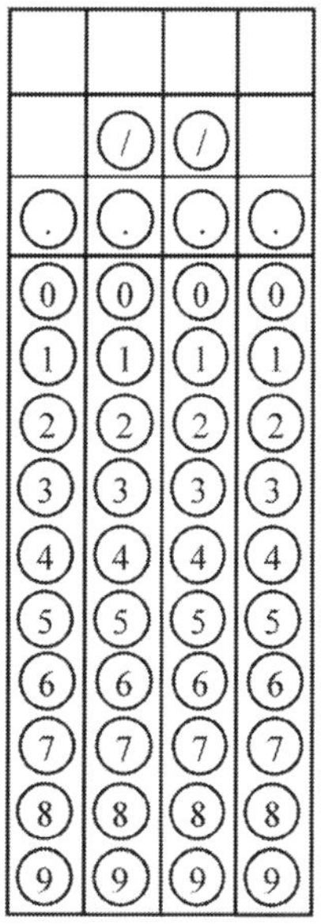

18. What number is equivalent to $16^{\frac{1}{4}} \times 16^{\frac{1}{2}}$?

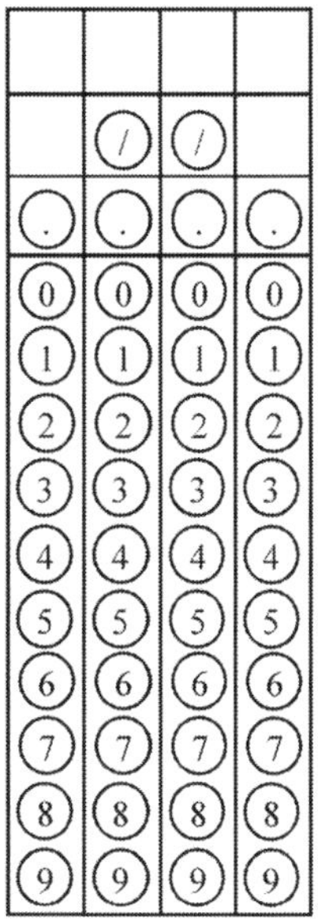

19. For what value of x is the following function undefined?

$$h(x) = \frac{6}{(x-9)^2 + 6(x-9) + 9}$$

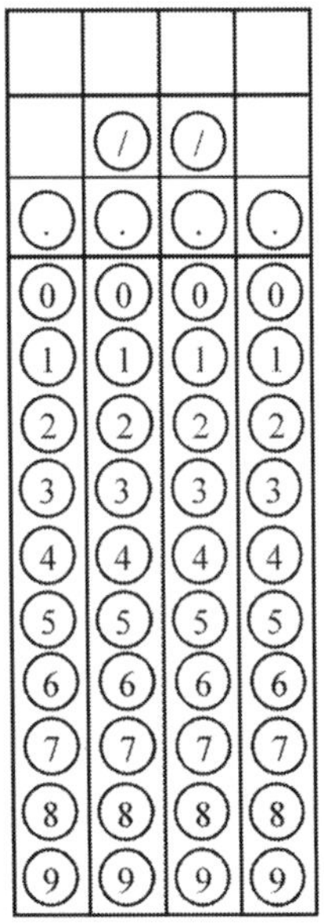

20. The following table shows values of two functions $f(x)$ and $g(x)$.

x	$f(x)$	$g(x)$
4	6	3
5	-8	5
6	2	-8

If $h(x) = -4 \times g(x) - f(x)$, what is the value of $h(6)$?

Calculator Questions

21. How many cases of cola can Lexi purchase if each case is \$3.50 and she has \$40?
   a. 10
   b. 12
   c. 11.4
   d. 11

22. A car manufacturer usually makes 15,412 SUVs, 25,815 station wagons, 50,412 sedans, 8,123 trucks, and 18,312 hybrids a month. About how many cars are manufactured each month?
   a. 120,000
   b. 200,000
   c. 300,000
   d. 12,000

23. Each year, a family goes to the grocery store every week and spends \$105. About how much does the family spend annually on groceries?
   a. \$10,000
   b. \$50,000
   c. \$500
   d. \$5,000

24. A grocery store sold 48 bags of apples in one day, and 9 of the bags contained Granny Smith apples. The rest contained Red Delicious apples. What is the ratio of bags of Granny Smith to bags of Red Delicious apples that were sold?

a. 48:9
b. 39:9
c. 9:48
d. 9:39

25. If Oscar's bank account totaled $4,000 in March and $4,900 in June, what was the rate of change in his bank account over those three months?

a. $900 a month
b. $300 a month
c. $4,900 a month
d. $100 a month

26. Erin and Katie work at the same ice cream shop. Together, they always work less than 21 hours a week. In a week, if Katie worked two times as many hours as Erin, how many hours did Erin work?

a. Less than 7 hours
b. Less than or equal to 7 hours
c. More than 7 hours
d. Less than 8 hours

27. What is the area of the following figure?

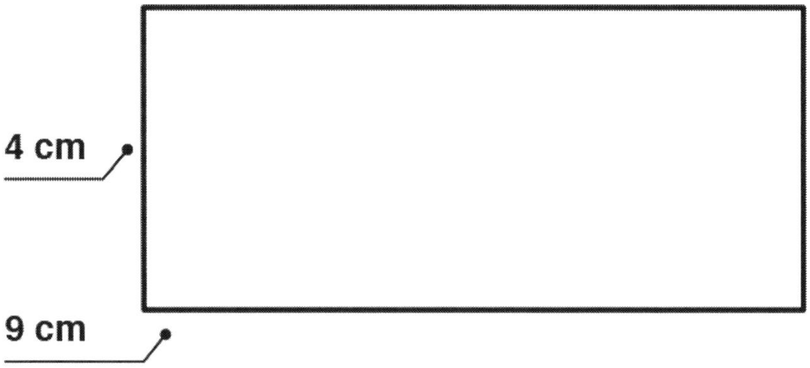

a. 26 cm
b. 36 cm
c. 13 cm^2
d. 36 cm^2

28. Which equation correctly shows how to find the surface area of a cylinder?

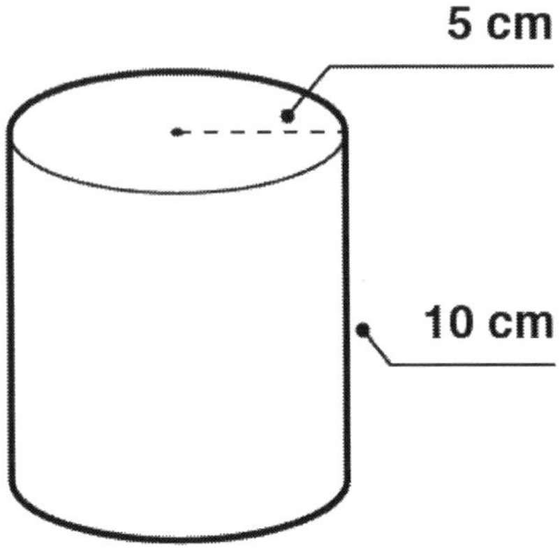

a. $SA = 2\pi \times 5 \times 10 + 2(\pi 5^2)$
b. $SA = 5 \times 2\pi \times 5$
c. $SA = 2\pi 5^2$
d. $SA = 2\pi \times 10 + \pi 5^2$

29. In order to estimate deer population in a forest, biologists obtained a sample of deer in that forest and tagged each one of them. The sample had 300 deer in total. They returned a week later and harmlessly captured 400 deer, and found that 5 were tagged. Using this information, which of the following is the best estimate of the total number of deer in the forest?

a. 24,000 deer
b. 30,000 deer
c. 40,000 deer
d. 100,000 deer

30. What is the correct factorization of the following binomial?

$$2y^3 - 128$$

a. $2(y + 8)(y - 8)$
b. $2(y - 4)(y^2 + 4y + 16)$
c. $2(y - 4)(y + 4)^2$
d. $2(y - 4)^3$

31. What is the simplified form of $(4y^3)^4(3y^7)^2$?

a. $12y^{26}$
b. $2{,}304y^{16}$
c. $12y^{14}$
d. $2{,}304y^{26}$

32. The number of members of the House of Representatives varies directly with the total population in a state. If the state of New York has 19,800,000 residents and has 27 total representatives, how many should Ohio have with a population of 11,800,000?

a. 10
b. 16
c. 11
d. 5

33. The following set represents the test scores from a university class: {35, 79, 80, 87, 87, 90, 92, 95, 95, 98, 99}. If the outlier is removed from this set, which of the following is TRUE?

a. The mean and the median will decrease.
b. The mean and the median will increase.
c. The mean and the mode will increase.
d. The mean and the mode will decrease.

34. Eva Jane is practicing for an upcoming 5K run. She has recorded the following times (in minutes):

25, 18, 23, 28, 30, 22.5, 23, 33, 20

Use the above information to answer the next three questions to the closest minute. What is Eva Jane's mean time?

a. 26 minutes
b. 19 minutes
c. 25 minutes
d. 23 minutes

35. What is the mode of Eva Jane's time?

a. 16 minutes
b. 20 minutes
c. 23 minutes
d. 33 minutes

36. What is Eva Jane's median time?

a. 23 minutes
b. 17 minutes
c. 28 minutes
d. 19 minutes

37. Use the graph below entitled "Projected Temperatures for Tomorrow's Winter Storm" to answer the question.

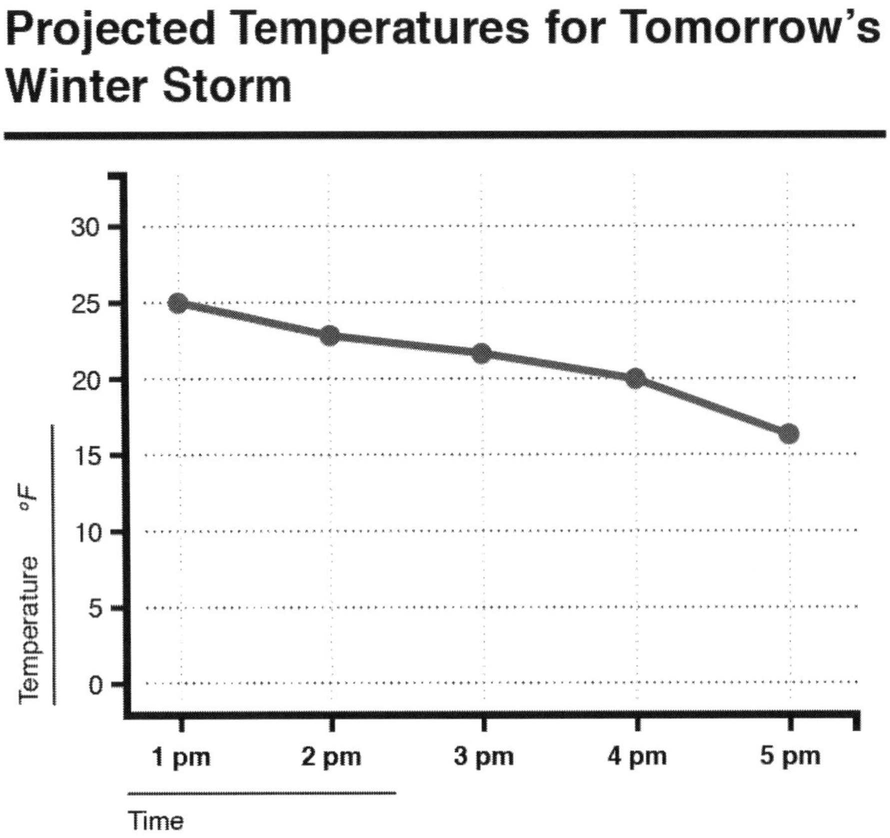

What is the expected temperature at 3:00 p.m.?

a. 25 degrees
b. 22 degrees
c. 20 degrees
d. 16 degrees

38. The function $f(x) = 3.1x + 240$ models the total U.S. population, in millions, x years after the year 1980. Use this function to answer the following question: What is the total U.S. population in 2011? Round to the nearest million.

a. 336 people
b. 336 million people
c. 6,474 people
d. 647 million people

39. What are the zeros of the following quadratic function?

$$f(x) = 2x^2 - 12x + 16$$

a. $x = 2$ and $x = 4$
b. $x = 8$ and $x = 2$
c. $x = 2$ and $x = 0$
d. $x = 0$ and $x = 4$

40. Which of the following values of x satisfy the following equation? Round your answer to the nearest hundredth place.

$$\frac{4}{3}\left(x - \frac{11}{15}\right) = 20$$

a. 15.73
b. 17.74
c. 10.73
d. 11.74

41. A pen costs $0.50 less than three times the cost of a pencil. If the pen and pencil together cost $5.10, how much more does the pen cost than the pencil?

a. $3.70
b. $4.25
c. $2.30
d. $1.40

42. The graph of a line passes through the points (5.72, 3.95) and (7.12, 6.75). Which of the following is the equation of the line?

a. $y = 2x + 9.67$
b. $y = 2x - 7.49$
c. $y = x - 11.44$
d. $y = \frac{1}{2}x + 3.95$

43. If $4x + 4y = 24$ and $5x - 6y = 8$, what is the value of $\frac{x}{y}$?

a. 2
b. 4
c. $\frac{1}{2}$
d. $\frac{3}{2}$

44. A line passes through the origin, the point $(a, -50)$, and $(-2, a)$. Which of the following could be the value of a?

a. 100
b. 50
c. 10
d. 5

45. Which of the following is NOT a solution to the inequality below?

$$6.5c - 11.2 \geq 9 + 8.5c$$

a. -10.1
b. -9.9
c. -10
d. -10.3

46. The number of online students at a university between 2015 to 2020 is 15 times the number of online students at the same university between 2010 and 2015. If the university had 2400 online students between 2015 and 2020 and x online students between 2010 and 2015, which of the following equations is true?

a. $\frac{x}{15} = 2400$

b. $x = 160$

c. $x + 15 = 2400$

d. $2400x = 15$

47. If the system of equations below has no solution and c and d are constants, what is the value of $|c - d|$?

$$-25 = cy + 48x$$

$$11 + 12dx = 10y$$

a. 14
b. 10
c. 6
d. 12

48. Debbie is taking an exam. The multiple-choice questions are worth 2.5 points and the short answer questions are worth 4 points. If she answers 15 multiple-choice questions correctly and x short answer questions correctly, which of the following expressions represents her total score on the exam?

a. $15x + 2.5$
b. $2.5x + 60$
c. $4x + 37.5$
d. 41.5

49. Which expression is equivalent to the following function?

$$g(x) = (4.2x - 3.2)^2 - (0.54x^2 - 1.21x + 11.81)$$

a. $4.2x^2 + 1.21x - 11.81$
b. $17.64x^2 - 28.09x + 10.24$
c. $17.1x^2 - 25.67x - 1.57$
d. $18.18x^2 - 12.23x + 22.05$

50. Consider the equation below. If $x > 0$, which is the correct value of x?

$$\sqrt{4x^2 - 12} = \sqrt{1012}$$

a. 256
b. 16
c. 26
d. -16

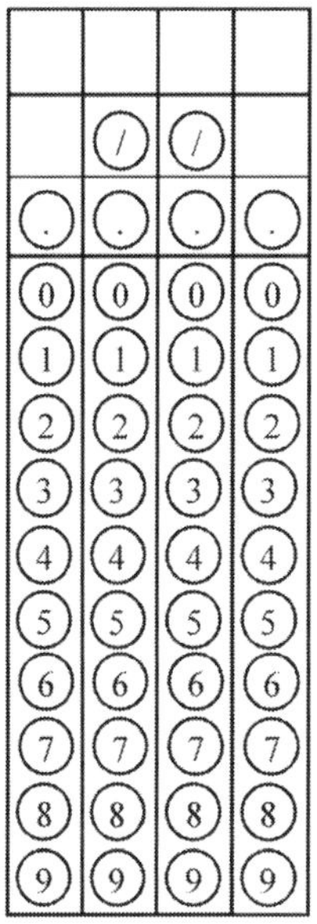

51. Gina took an algebra test last Friday. There were 35 questions, and she answered 60% of them correctly. How many correct answers did she have?

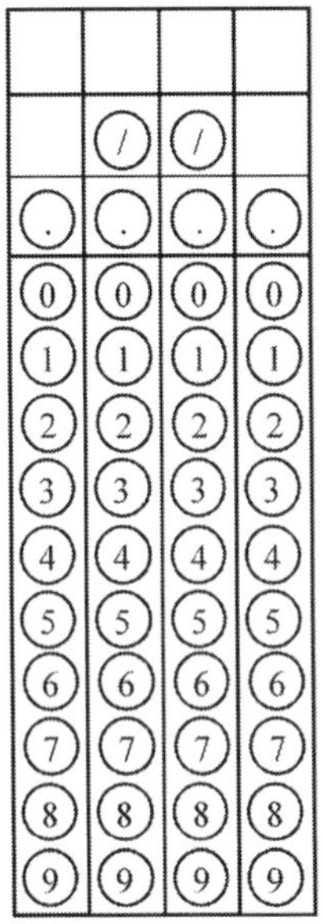

52. Paul took a written driving test, and he got 12 of the questions correct. If he answered 75% of the questions correctly, how many problems were there in the test?

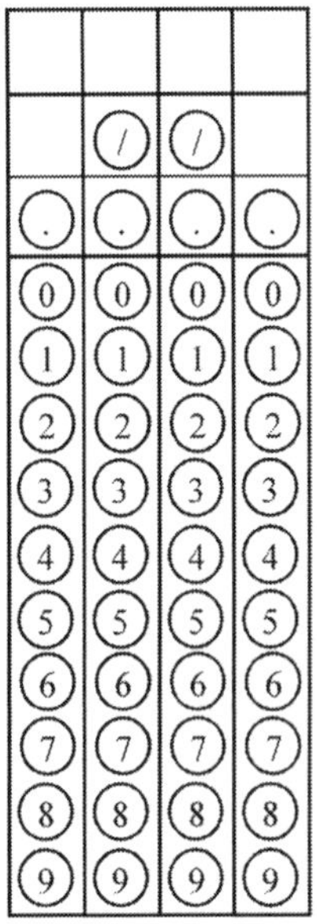

53. Bindee is having a barbeque on Sunday and needs 12 packets of ketchup for every 5 guests. If 60 guests are coming, how many packets of ketchup should she buy?

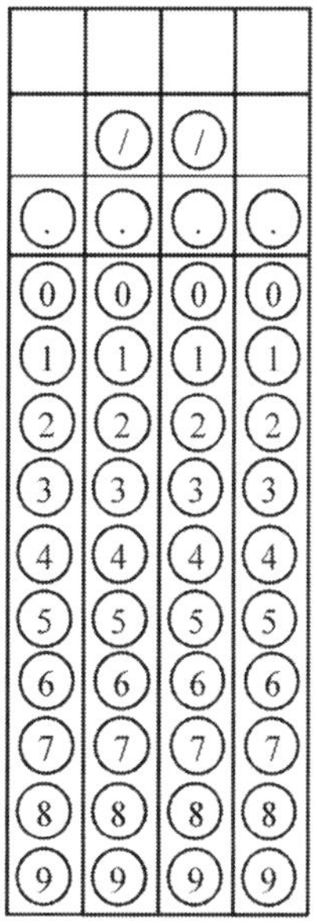

54. How many centimeters are in 3 feet? Round your answer to the nearest tenth of an inch. (Note: $2.54 \text{ cm} = 1 \text{ in}$)

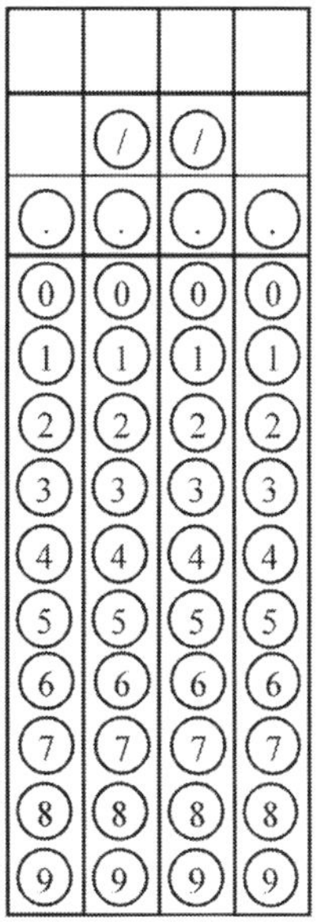

55. Triple the difference of five and a number is equal to the sum of that number and 5. What is the number?

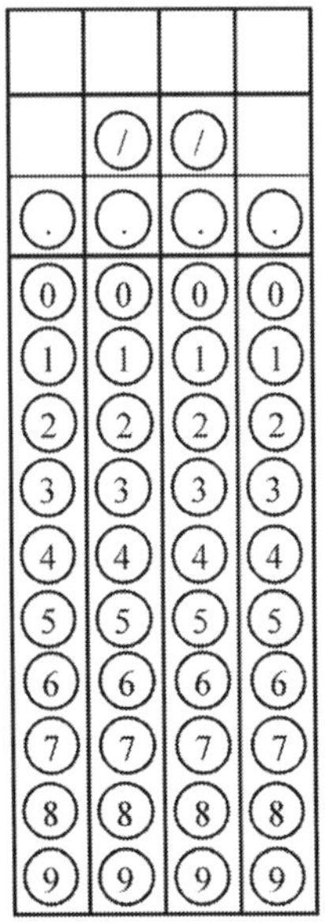

56. What is the perimeter of the following figure in meters?

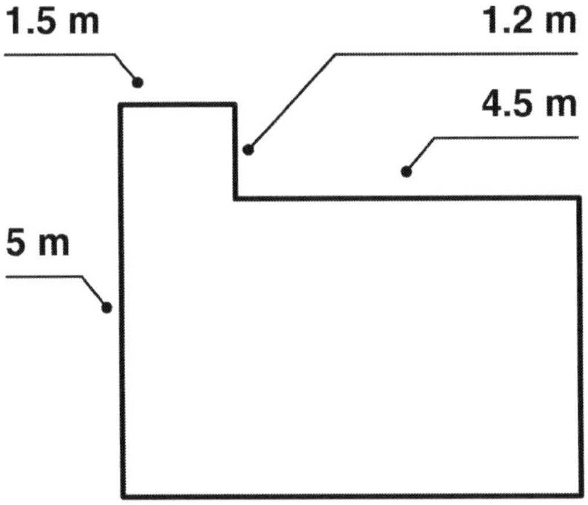

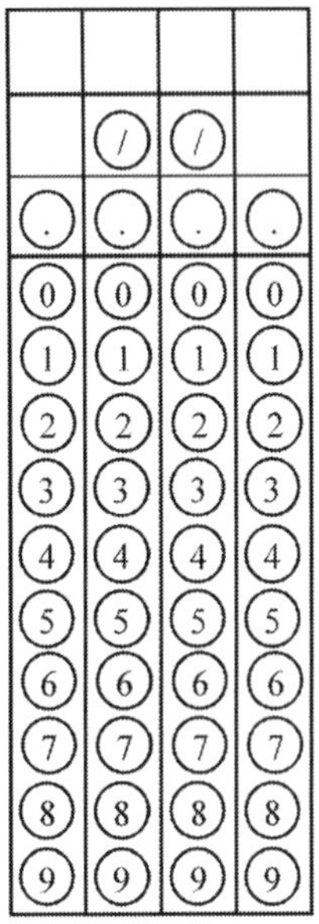

57. If the line of the following equation is graphed in the xy-plane, what is the y-intercept of the line?

$$\frac{4}{3}x - \frac{11}{9}y = -55$$

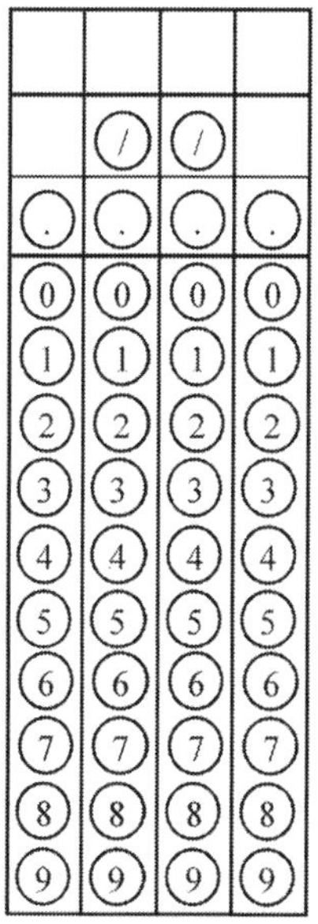

58. If $8.75x = 7 - 0.25b$, what is the value of $b + 35x$?

Writing Prompt

Read the passage below. Take into consideration how the author uses evidence to support their claims. Consider the following:

- Evidence to support claims, such as facts or examples
- Logical reasoning to develop main ideas and connect claims and evidence
- Stylistic elements, such as how figurative language appeals to the readers' emotions

Write an essay explaining how the author of the following passage builds their argument. Examine how they use one or more of the features above to enhance the logic and persuasiveness of their claim. Remember your job isn't to agree or disagree with the author, but to analyze the way in which they build their argument.

> Florentine painting between Giotto and Michelangelo contains the names of such artists as Orcagna, Masaccio, Fra Filippo, Pollaiuolo, Verrocchio, Leonardo, and Botticelli. Put beside these the greatest names in Venetian art, the Vivarini, the Bellini, Giorgione, Titian, and Tintoret. The difference is striking. The significance of the Venetian names is exhausted with their significance as painters. Not so with the Florentines. Forget that they were painters, they remain great sculptors; forget that they were sculptors, and still they remain architects, poets, and even men of science. They left no form of expression untried, and to none could they say, "This will perfectly convey my meaning." Painting, therefore, offers but a partial and not always the most adequate manifestation of their personality, and we feel the artist as greater than his work, and the man as soaring above the artist.

From *The Florentine Painters of the Renaissance* by Bernhard Berenson

Answer Explanations #2

Reading

1. A: The point of view of the narrator of the passage can best be described as third-person omniscient. The narrator refers to the characters in the story by third-person pronouns: *he, it,* and *they.* The narrator also comes across as "all-knowing" (omniscient) by relating the information and feelings about all the characters (instead of just those of a single character). Second-person point of view would incorporate the second-person pronoun, *you.* First-person point of view would incorporate first-person pronouns: *I* and *we.* Finally, the third-person objective point of view would also refer to the characters in the story by third-person pronouns *he, it,* and *they.* However, the narrator would stay detached, only telling the story and not expressing what the characters feel and think.

2. B: The chronological order of these four events in the passage is as follows:

1. The servants joyfully cheered for the family.
2. Mrs. Dashwood discussed improvements to the cottage.
3. Elinor hung her drawings up in the sitting room.
4. Sir John Middleton stopped by for a visit.

3. C: In the course of this passage, the Dashwoods' attitude shifts. The narrator initially describes the family's melancholy disposition at the start of their journey from Norland. However, the narrator then describes their spirits beginning to lift and becoming more upbeat as they make their way through the scenic English countryside to their new home at Barton Cottage. Although the remaining answer choices may contain some partial truths (appreciation for the family's former privileged life and fear about the family's expedition), they do not accurately depict the narrator's theme throughout the entire passage. For example, the family did not express disdain for their new landlord or doubt their future ability to survive.

4. D: Reality versus expectations is a theme of this passage. Although Mrs. Dashwood expects her new financial circumstances to be trying, and the family expects to miss their former place of residence for quite some time, the Dashwoods seem to begin to adapt well with the help of Sir John Middleton, their generous new landlord. A theme of love conquers all is typically used in literature when a character overcomes an obstacle due to his or her love for someone. A theme of power of wealth is used in literature to show that money either accomplishes things or is the root of evil. Finally, a theme of wisdom of experience is typically used in literature to show that improved judgment comes with age.

5. A: The statement at the start of paragraph five in this passage signifies the family is going through a transition. The Dashwoods find themselves in a time of great transition as they learn to accept their family's demotion in social standing and their reduced income. The statement was not meant to signify that Mrs. Dashwood needs to obtain meaningful employment, that the family is going through a growth phase, or that the Dashwood children need to be concerned about the future.

6. B: The new landlord is a cousin to the Dashwoods. This accurately depicts their relationship. The new landlord was not a former social acquaintance, Mrs. Dashwood's father, or a long-time friend of the family.

7. C: The narrator describes the generosity of Sir John Middleton to explain how his kindness eased the family's adaptation to their new home and circumstances. He played a very important role in making the Dashwoods feel comfortable moving to Barton Cottage. The narrator did not describe Sir John Middleton's generosity to identify one of the many positive traits that a landlord should possess, to explain how a landlord should conduct himself in order to be successful, or to demonstrate that he did not need to be cold and businesslike all the time.

8. D: The narrator's reference to the Dashwood's new home as being "defective" means that the building's look and feel do not resemble that of a typical cottage. The narrator describes the building as being regular, with a tiled roof (rather than a thatched roof), with shutters that are not green, and with walls not covered with honeysuckles—all characteristics that a traditional cottage would possess. Although the other three answer choices are true statements about their new home and property, they do not relate to the narrator's statement that the new home is "defective."

9. C: The overall tone of the passage is upbeat. Although the passage begins with the family in a melancholy disposition at the start of their journey (and perhaps a bit apprehensive), their spirits begin to lift as they make their way through the scenic English countryside to Barton Cottage.

10. B: The narrator is describing how exuberantly Sir John Middleton expresses his pleasure at gaining the Dashwoods as tenants. Sir John Middleton is overjoyed to welcome his cousins to the cottage on his property, which he expresses through his words, gifts, and invitations to dine with him. He just goes a bit too far in demonstrating his delight, for which the Dashwoods forgive him. Sir John Middleton does not express any fake friendliness. The phrase was not used to demonstrate the language from the time period in which the piece was penned. Finally, Sir John Middleton's offer for the Dashwoods to stay at Barton Cottage is indeed genuine.

11. A: The author would agree that an advantage associated with the division of labor proposed by Adam Smith is that dexterity increases as employees complete repeated tasks. The remaining three answer choices are known disadvantages related to the division of labor concept.

12. B: The author mentions that the "first great builder of iron bridges was a stone-mason" in paragraph five in order to demonstrate how the third advantage associated with the division of labor proposed by Adam Smith is not entirely true. Even though the author agrees with Adam Smith that the invention of machines helps employees to save time, he does not agree that employees who are intimately involved in a daily routine of work come up with the ideas for those inventions (i.e., "the inventor of the power-loom was a clergyman," not a weaver). The remaining answer choices are not addressed by the author in the passage.

13. C: The passage's main idea is best expressed by the answer choice "advantages associated with the division of labor." The passage does not focus solely on the effect of specialization of labor, which consists of increasing productivity by dividing up larger tasks into smaller tasks to be completed by workers with specialized skills.. The passage is not concerned with disproving Adam Smith's economic principles. Finally, macroeconomics deals with the larger economy as a whole and topics such as interest rates and gross domestic product. Thus, macroeconomics is not discussed in this passage.

14. D: In paragraph six, the author lists classification of employees by their abilities as the second most important advantage associated with the division of labor. The author states that this is "the greatest advantage (next to the dexterity of the workmen)," which is not mentioned as an advantage by Adam Smith. The remaining three answer choices are listed in the passage as advantages associated with the division of labor. However, they are not listed by the author as being the second most important advantage.

15. A: The author uses the word "bodily" in the second paragraph to mean tangible. The author is referring to the hands-on acts of dancing, gymnastics, and playing musical instruments. The remaining three answer choices are also meanings of the word *bodily*. However, they do not fit the author's use of the word *bodily* in this passage.

16. B: The author's attitude toward Adam Smith in this passage can be best described as respectful. The author does not adore Adam Smith, nor does he question or dismiss him and his theory. However, the author expands on the advantages associated with the division of labor to add in his own points where appropriate, and he describes one additional advantage not mentioned by Adam Smith.

17. C: Based on the second advantage, the author would suggest the most efficient course of action for an employee working in a copy center would be to produce black and white copies for a few hours and then change their focus to shipping out copy orders to customers. This is where the author says that he differs from Adam Smith. He believes that there is a benefit to having an employee gain relief from experiencing a change of occupation (or tasks) throughout their work day. Therefore, the author would not suggest that the most efficient course of action is having an employee focus on only a single task (as Adam Smith believed). Completing tasks in different locations is an idea that both the author and Adam Smith would not agree with. They were both proponents of performing tasks in one same physical location to cut down on transfer time. Finally, the author does not mention granting an employee unlimited flexibility in scheduling their work tasks.

18. D: This passage supports the idea of milking machines being used to help speed up the retrieval of milk from dairy farms. This is related to the third advantage: inventions tend to save labor. The passage does not mention that repetitive tasks may lead to stability in employees' work. Although the author mentions there is a benefit to having an employee gain relief from experiencing a change of occupation (or tasks) throughout their work day, he does not state that one employee should complete all the tasks in a workflow. Finally, the passage does not mention the downside of inventions, which is that sometimes they result in a loss of employee jobs.

19. A: Staffing the doll factory is an example of classing employees by their capacity; staff will be hired based on their skills for the tasks that are involved with manufacturing dolls. There is no mention of training employees on repetitive tasks to increase their dexterity or exclusively hiring only experienced employees. Finally, there is also no mention of introducing machines to assist employees with unfamiliar tasks.

20. B: The author uses the phrase "animal spirits" in paragraph three to explain the positive effect that variety in work has on employees. This is one of the areas in which the author expands on one of Adam Smith's advantages of the division of labor. Adam Smith says that time can be saved if employees focus on one type of work. The author is not dismissing his claim. However, he mentions that some employees may instead benefit from changing the type of work that they do during the day, and such variety invigorates some employees' animal spirits. The remaining answer choices (working beyond fatigue, failing to receive proper training, and employees being adversely affected when they change tasks) are not mentioned in the passage.

21. A: Providing money to allow a student the opportunity to attend an educational workshop is the simplest way to do something benevolent, according to the author of passage A. In the second paragraph of the passage, the author states that pecuniary (or monetary) aid is the easiest form of benevolence and that it requires the least amount of sacrifice. The remaining answer choices would all involve much greater amounts of sacrifice.

22. B: The phrase "mental exercise" is used in paragraph two of passage A to represent trying to see things from another person's point of view. In paragraph two, the author is discussing practicing benevolence by placing oneself in the situations of others. The author does not make mention of a person only taking their own perspective into consideration, improving an individual's memory recall, or strengthening a person's concentration.

23. C: In passage A, the term "heads" means manner. The author is discussing treating the practice of benevolent affections in almost the same way or manner as when considering the principles of justice. The remaining answer choices (discord, unlikeness, and opposition) do not make sense in this context.

24. D: The author of passage A states that justice can be exercised at the expense of benevolence, but this premise is not true the other way around ("benevolence is not to be exercised at the expense of justice"). Additionally, acts of benevolence require selflessness (not acts of justice). Finally, benevolence cannot take the place of the payment of debts that are owed per acts of justice.

25. A: The author of passage B says that giving food and money to beggars is seen as a weakness because that type of charity encourages laziness and corruption ("giving alms to common beggars" leads to an increase in "idleness and debauchery"). Although the remaining answer choices may indeed be true, they are not mentioned by the author of passage B.

26. B: The phrase "their gentle dominion over the breasts of men" refers to the social virtues, compassion and benevolence. In this paragraph, the author discusses how the goal of benevolence is to promote the happiness of society. The other answer choices are mentioned in paragraph five, but the author does not propose they rule over people's inner spirits ("the breasts of men").

27. C: The author's attitude toward luxury in passage B can best be described as objective. The author is impartial and simply represents facts and attitudes about luxury. The author does not say luxury is unworthy of consideration (dismissive), does not think romantically or sentimentally about luxury (nostalgic), and does not worship luxury (adoring).

28. D: Both passage A and passage B are written in first-person point of view. The author uses the first-person plural pronouns *we* and *us*. Second-person point of view would use the second-person pronoun, *you*. Third-person objective point of view would utilize third-person pronouns: *he, she, it,* or *they,* and the author would stay detached by telling the story. Finally, third-person limited omniscient view would also use the third-person pronouns, and the author would tell the entire story through the eyes of a single character.

29. A: The tone of passage A is objective, and the tone of passage B is earnest. The author is passage A is uninfluenced by emotion, while the author of passage B shows deeper feelings with the incorporation of words in all capital letters and the phrases and examples which he selected. A cynical tone is used when an author is mocking others, and a sarcastic tone is used when an author wishes to state a negative opinion in an ironic way. Finally, an excited tone is used when an author is very happy about events, and an ambivalent tone is used when an author is unsure or undecided about something.

30. B: Compared to the author of passage A, the author of passage B stresses the importance of exercising benevolence for the purposes of contributing to the overall happiness of society ("bestow happiness on human society"). The author of passage A mentions the importance of using benevolence to alleviate the distresses of others and to protect the reputations of others. The concept of attracting karma is not mentioned in either passage.

31. A: The main focus of the passage is a discussion of the causes, symptoms, and treatments associated with the various types of tenosynovitis. The remaining answer choices are all mentioned throughout the passage; however, they are separate, subordinate topics, not the main idea of the passage.

32. B: Placing the tendon at rest and possibly utilizing a splinting device is mentioned in the passage as a treatment for tenosynovitis crepitans of the ankle. Depositing urate of soda under the endothelial tendon covering is used to treat gouty tenosynovitis. Inducing hyperemia and making small incisions to release pus, as well as incorporating active and passive movements to prevent tendon adhesions, are both mentioned in the passage as treatments for suppurative tenosynovitis.

33. C: The statement is supported by the information found in the third paragraph, since the author states that only a small proportion of individuals who overuse their tendons are afflicted with the condition. This means that something else, such as the toxic factor that is mentioned, must be the reason.

34. D: According to the author, a patient having a sensation of crepitation or friction when the tendon is moved in its sheath would be presenting with a case of tenosynovitis crepitans. Visible nodular swellings are a clinical feature of gouty tenosynovitis. Finally, cellulitis and finger wound infections are both clinical features of suppurative tenosynovitis.

35. A: The authors use first-person point of view in this passage, which can be determined by the use of the first-person pronoun, *we*, in paragraph eight. Second-person point of view would involve the authors utilizing the second-person pronoun, *you*. Third-person limited omniscient view would involve the authors using third-person pronouns (*he, she, it*, or *they*), and the authors would tell the entire story through the eyes of a single character. Finally, third-person objective point of view would also involve the authors using the third-person pronouns: *he, she, it*, or *they*. However, the authors would stay detached and would not refer to themselves with the pronouns "I" or "we."

36. B: Pus spreading to the adjacent bones and joints of the wrist may be the end result of an infected wound of the finger, according to the passage. Based on the options provided, this answer choice would have the most finality for an individual suffering with this condition. The other answer choices are all complications that can result from an infected wound of the finger, but Choices *A* and *C* are not as serious as the infection spreading from the finger to the bones and joints of the wrist. Choice *C*, amputation of the finger, is explicitly mention has a condition that can *precede* suppurative tenosynovitis, so it is not a good answer choice for an end result.

37. C: The phrases are used in paragraph four to describe the friction associated with tendon movement in the sheath. The authors utilize the phrases to express how the friction can present in varying degrees. Milder friction can be "soft like the friction of snow"; more intense friction can "resemble the creaking of new leather." The other answer choices are simply additional features of tenosynovitis crepitans.

38. D: The authors state in paragraph six that tenosynovitis crepitans at the ankle can result from the pressure of ill-fitting boots and that the individual affected would experience pain when raising his or her heels from the ground. Therefore, based on this passage, if a construction worker wearing tight boots experiences pain when raising his heel, he will more than likely be diagnosed with tenosynovitis crepitans.

39. A: The authors included the words "washerwomen" and "riveters" in paragraph five to give examples of workers whose repetitive motions on the job can result in tenosynovitis crepitans. The authors make no mention in the passage of the importance of healthcare for blue-collar workers or ergonomics in the workplace. Finally, the authors mention that recurrent attacks of inflammation are likely to occur in individuals suffering from gouty tenosynovitis. However, they do not provide any examples related to this in the passage.

40. B: Based on this passage, if a woman presents to her doctor with a nodular swelling on her thumb the size of a blueberry, she will more than likely be diagnosed with gouty tenosynovitis. This is the only affliction the authors mention is characterized by the formation of visible nodular swellings that vary in size from a pea to a cherry.

41. C: The main idea of the passage is Choice *C*: Rural architecture in America is inconsistent in style and construct and is thus offensive to the farmer's need of symmetry and order. The first paragraph explains the inconsistency of rural architecture. The second paragraph expands on this point and acknowledges that better architecture exists in the cities. The third paragraph concludes how the farmer's need for symmetry and order is ignored in rural architecture.

42. B: The author's tone in the passage is *critical*, Choice *B*. We see the author criticizing and offering their opinion on the current state of rural architecture. There is a mix of passion and neutrality in the author's tone, but neither one wins out, so Choices *A* and *C* are not the best answers. *Joyous* is also incorrect, as there is more of a critical tone than a tone of joy.

43. A: The meaning of the word *incongruous* means absurd, out of place, or unequal. The author mentions the style of the farmhouses being "incongruent" and remarks that there is an "absence of method" to their construction.

44. A: The author says that human beings should leave the world better than when they found it. This is found in the second paragraph toward the bottom: "Every human being is bound, under ordinary circumstances, to leave the world somewhat better, so far as his own acts or exertions are concerned, than he found it."

45. D: The author implies that America's rural architecture is poor and uncivilized. The author says in the third paragraph that rural architecture is "discordant in appearance" and also "uncouth," which means awkward or uncultivated.

46. C: The word *discordant* means conflicting, or *not* in harmony. The author uses this word to describe the rural architecture, meaning that it is inconsistent in style and clashes with any kind of logical construction.

47. B: The main purpose of this passage is to explain the definition and etymology of quilts. The second and third paragraph describes the etymology and definition of *quilt* from two different sources.

48. D: The author uses direct quotes in this passage to provide evidence. Specifically, they use direct quotes from two different encyclopedia sources. Direct quotes are a great way to provide evidence to a text because you can incorporate them into your own sentence, and they add credibility if they are from a trusted source.

49. C: The author says that the Encyclopedia Britannica gives a more specific definition than Webster. This is found in the third paragraph where it says, "The 'Encyclopædia Britannica' is a little more explicit," which means that it's more specific than Webster.

50. C: Comforts are blankets that are stitched together that are thicker than quilts. This is found in the very last paragraph when the author says, "In the United States the distinction has been carried even farther and a quilt is understood to be a light weight, closely stitched bedcover. When made thicker, and consequently warmer, it is called a "comfort." This describes a separate blanket than the quit called a "comfort," today known as "comforters."

51. B: According to the passage, the words *counterpane* and *quilt* come from the same origins. This is found in the third paragraph: "Thus 'counterpane,' a coverlet for a bed, and 'quilt' are by origin the same word."

52. A: This passage is informative, because it presents information to the audience about the origin and etymology of *quilt*. Choice *B*, persuasive, is incorrect, because the author does not try to persuade the audience to think or act in any way. Choice *C*, narrative, is incorrect, because a story is not being told. Choice *D*, cause and effect, is incorrect, because this sort of relationship is not shown in the passage.

Writing and Language

1. B: The underlined portion of this sentence is a preposition. Prepositions connect an object to nouns, to pronouns, or to a phrase representing a noun. Here, the noun is "view," Hawthorne's tragic view, and the object is "the literature of the United States." The preposition is the word "in," which indicates the vision is already within the body of literature. But the beginning of the sentence makes it clear that Hawthorne's vision is a new addition, so the correct preposition is "to," to show possession (the literature now includes Hawthorne's vision). "From" is incorrect because it is the opposite of "to," and "with" is incorrect because it signals that Hawthorne and the literature are co-contributors.

2. C: Collective nouns denote a group as a single unit, and they typically take a singular pronoun ("Mankind had its finest hour.") In this passage, "man" is taken as the universal, collective subject. "His" agrees with "man" and in this context is grammatically correct. Choice *B* ("her") would make the valid point that traditional constructs unfairly exclude women, but that argument is irrelevant to the discussion of Hawthorne's allegory of the human heart. Choice *D*, people's, is grammatically correct, but it is not the best choice, because the sentence is already about one collective subject, "man," and does not need another collective term, "people."

3. D: Choice *D* is a singular possessive noun. Choice *B* is the plural possessive of "humanities," the branches of learning. Choice *C*, "humanitys," is not a word.

4. A: The highlighted portion is an example of two independent clauses held together by a semicolon. The first clause is "but a few feet in, the light dims and warmth turns to chill," and the second one is "the visitor stumbles first in confusion, then in terror." The introductory phrase "but a few feet in" is an absolute phrase and must end with a comma. Inserting a comma after "dims" is inappropriate because the conjunction "and" connects only two actions. In the second clause, the comma after "confusion" substitutes for the conjunction "and." Choice *B* is incorrect because the comma separates the noun "light" from its verb "dims." Choice *C* is incorrect because the commas are needed to separate the phrases within the clauses. Choice *D* is incorrect because it lacks the appropriate commas and because the clause following the colon does not introduce a summary or an explanation.

5. D: "Farther" indicates additional distance, whereas "further" indicates an additional amount of something abstract, such as time. Choice *B* interrupts the sense of journey the writer is describing, and therefore is not the best answer. Choice *C* is awkward; "toward" rather than "to" is correct.

6. C: Choice *C* is correct because a colon is used to introduce a summary or an explanation. The phrase "the beauty that lies beyond fear and hopelessness" further describes Hawthorne's vision of human nature. Choice *B* is incorrect because it is a fragment, not a complete sentence. Choice *D* is incorrect because parentheses are used to enclose nonessential elements such as minor digressions.

7. B: Choice *B* is correct because inserting the article "an" creates parallel construction with "a natural morality." Choices *C* and *D* are incorrect because they lack parallel construction.

8. D: The word "affect" is a verb that means "to have an influence on," whereas "effect" in this sentence is a noun synonymous with "result." Choices *B* and *C* are incorrect because they alter the writer's intended meaning.

9. B: While Choices *A*, *C*, and *D* might all describe the moral setting of New England's Puritan community, Choice *B* is the closest synonym to "rigid."

10. C: The correct answer is Choice *C* because the phrase "which took place between 1760 and 1840" is set off with commas. Choice *B* is incorrect because it lacks a comma after "1840," and Choice *D* is incorrect because the comma after "place" interrupts the phrase "took place between 1760 and 1840."

11. D: Choice *B* is incorrect because a word is missing: that. The phrase should be "problems *that* threatened European society." Choice *C* is incorrect because it is awkward and confusing. Choice *D* is the correct answer because it uses "that," which does not take a preceding comma, as "which" must.

12. D: Choice *D*, "transformed," is the most accurate substitution because the essay describes the dramatic demographic changes that happened to European society during the Industrial Revolution. Choice *A* ("transfigured") suggests a radical change in figure or appearance, and as such is less exact than "transformed." Choice *B* ("disrupted") does not convey the idea of change, and Choice *C* ("destroyed") is inaccurate because the existing social order was not destroyed.

13. C: The correct answer is Choice *C* because it includes the missing word "as" ("digging *as* deeply"). Choices *A*, *B*, and *D* are incorrect because they do not include the missing word.

14. A: Choice *B*, "indelible," is incorrect because it means permanent. Although industrialization did become permanent, the paragraph is outlining the development of industrialization, not proving its permanence. Choice *C* ("inadvertent") erroneously implies that industrialization was accidental or unintentional. Choice *D* ("inadvisable") classifies industrialization as "unwise," and that thought is not in line with this story of how industrialization developed.

15. A: Choice *A* is the best selection, because it provides the link from the expense of building factories to the capitalists, which led to industry becoming the new source of wealth. Choice *B* is awkwardly worded, and although it correctly states the need for capitalists, it does not put them into action. Choice *C* is factually correct, but is not the best choice because the tone is more casual than that of the rest of the essay. Choice *D* is factually incorrect and contradicts information supplied earlier in the paragraph: supply, demand, and markets existed before the Industrial Revolution.

16. B: Choice *B* is the clearest rewrite. Forming two sentences allows for linear sequencing of the activities.

17. A: Choice *A* is the correct answer. "Amount" refers to items that cannot be counted, whereas "number" refers to something, such as people, that can be counted.

18. C: Choice *C* is the correct answer because it uses past tense ("had"), which is consistent with the rest of the paragraph.

19. B: While cardiologists (Choice *C*) and the American Obesity Association (Choice *D*) are actively dealing with the obesity epidemic in the United States, this paragraph focuses on a hospital's response. The correct answer is Choice *B*, "facilities." Choice *A* is incorrect because the closest definition of "faculties" that might fit in this context is "teachers and instructors in branches of learning in a college or university." If the writer had intended to use "faculties," they would have identified which branch, such as "medical faculties."

20. A: The spelling and punctuation are correct as written (Choice *A*). Choice *B* lacks a comma between the town and the state, as well as a comma after the state, to enclose the clause. In Choice *C*, Massachusetts is misspelled, and if Choice *D* were used, the essay would then contain less factual information than the original.

21. B: "BMI's" is incorrect because it is the possessive form of "BMI." Choice *B* ("BMIs) is correct because it is the plural form of "BMI." Choices *C* and *D* are incorrect because the acronym is spelled out in full; as it has already been introduced to the reader, it should be used here, not the full term.

22: C: Choice *C* is the correct answer because it states that sleep apnea *can* cause problems; the other choices imply that it always causes problems. In addition, Choices *B* and *D* each have an additional error. Choice *B* is written in the incorrect tense—it uses past tense instead of present tense. Choice *D* omits mentioning that the problems are caused during surgery.

23. C: Choice *A* is incorrect because in the underlined phrase "patients" is plural, indicating more than one patient. Choice *B* is incorrect because "patient" is positioned as a descriptive adjective, rather than as a noun. Choice *D* is partly correct because "patients'" is possessive, but it is incorrect because the word is in the plural form. Choice *C* is correct: it is the singular, possessive form of "patient."

24. A: This question involves proper placement of the adverb "only," which should precede the noun/verb combination it modifies, and it should also be as close to the combination as possible. The underlined phrase, as written, is correct. The device was not in the operating room on a regular basis; one was stocked if the staff knew in advance that the patient had a high BMI. The action being modified is the expectation of difficulty, therefore "only" precedes it and is as close as possible to the words "if difficulty was expected." Choice *B* is incorrect because the adverb follows the action statement. Choice *C* is incorrect because, by following directly behind "equipment," the word "only" attaches itself to that noun. Choice *D* is incorrect because the sentence does not make sense with that wording.

25. D: The correct answer is Choice *D*, "extended." Choices *A* ("unreasonable "), *B* ("outrageous "), and *C* ("unacceptable") are subjective and imply value judgments. "Extended," on the other hand, is factual and tells the reader specifically what it is about the amount of time that leads to rhabdomyolysis.

26. C: The comma before the conjunction "and" in Choice *A* makes this phrase grammatically incorrect. The same is true of Choice *B*. Choice *D* is incorrect because the dashes give the writing a jerky quality. Choice *C* is a simple statement of fact, does not contain a comma before the conjunction, and does not even need dashes.

27. A: A personal pronoun is needed in this phrase, and "their" (Choice *A*) is the correct one. As the reader is not told the patient's gender, the personal pronouns "his" and "her" would be incorrect, and "it" (Choice *D*) would not apply to a person. "They're" (Choice *B*) is a contraction of "they are," which does not make sense in this sentence. Choice *C* ("there") is an adverb, meaning "at, in, or to a place," which also makes no sense.

28. D: "Afflicts" is an acceptable substitute for "affects," in the context of this sentence. Choices *A* (destroys") and *C* ("kills") are both incorrect choices because not everyone who is affected by a stroke is killed by it. Choice *B* ("effects") is not correct because it is a noun.

29. D: The key phrase in the underlined passage is "*risk* of stroke." The demographics show who in the population is prone to stroke, but they don't guarantee that those people will have strokes. Likewise, they don't guarantee that anyone falling outside of the categories listed will not have a stroke. The correct answer is Choice *D*. It is possible that Carlos might have a stroke in his lifetime.

30. C: Inserted at the end of the paragraph, the sentence about pseudobulbar affect is confusing because the condition has not been introduced yet, and so the reader has no context to put the comment in. The paragraph ends without any additional information that would justify the sentence being there. Choice *A* is incorrect because the information might be very relevant to the essay—but only if it is included in the correct place. Choice *B* is incorrect because the additional information does not support the topic sentence, "What exactly is a stroke?" Choice *D* is incorrect not because the statistic is relatively positive, but because stroke survival is not the topic of the sentence.

31. B: The logical order is the order in which the events occur, as presented in Choice *B*. The blood flow disruption begins, it can be disrupted for up to twenty-four hours, and death can be a result.

32. C: Choice *C*, "poles," is the closest synonym because as with extremes, poles lie at either end of the spectrum. Choice *A* ("spectrums") is incorrect because there is only one spectrum and it is the continuum of all possibilities that lie between the extremes, including the extremes. Choices *B* ("states") and *D* ("positions") do not connote the sense of opposites that "extremes" does.

33. B: Choice *B* is correct because the comma separates the main clause in the sentence from the subordinating clause. Choice *C* would be correct if there was a comma after "brain" as well. Choice *D* is incorrect because the comma interrupts the prepositional phrase "to CNS tissue."

34. B: The correct answer is Choice *B*. The paragraph discusses clots and plaque that form in the CNS, and clots and plaque that form elsewhere. "Locally" indicates a position in the CNS itself, whereas "a blood clot or atherosclerotic plaque develops elsewhere, such as the heart" indicates a position outside of the CNS.

35. B: A semicolon separates main clauses that are not joined by a coordinating conjunction, such as "and." Choice *B* breaks this long sentence into two main clauses, each of which can stand on its own as a complete sentence.

36. C: The correct answer is Choice *C*, the semicolon. Because there is a set of dashes later in the sentence, Choice *B* ("dash") won't work because dashes are used only in pairs, and adding one would make three. Choice *A* ("slash") is used to separate two options, not two parts of a sentence. A period,

Choice *D*, would be appropriate only if the lowercase "t" in "there" appeared as a capital "T" in the sentence.

37. D: "Represented" is most accurate word. "Marked" (Choice *A*) describes what the triplets, triads, and trios do, rather than what they are—they mark, or flag, the concept of three-ness. Choice *B* ("earmarked") means "marked with an identifying symbol" and is an interesting choice, but not one that is as synonymous as "represented." Choice *C* ("included") is not grammatically correct.

38. B: "The notion of" should be deleted, because "The significance of three-ness..." is much more direct. "Of these cultures" (Choice *C*) is a necessary reference to the Indo-European societies. "Numerous" (Choice *D*) attests to the quantity of these events, and it is stronger than simply "abounds with events."

39. C: The error lies in Choice *C*. "Their" is the correct spelling of the possessive form of "they."

40. B: "Clearly and uniformly" is the closest substitute. Choice *A*, "discerningly and constantly," bears little similarity to the original pair aside from beginning with the same letters. Choice *C*, "definitely and evenly," is not immediately clear and would require explanation. In Choice *D*, "easily" is not a synonym of "distinctly."

41. A: "Dwellers" is the closest synonym to "inhabitants." Both are plural nouns and mean "people who reside in or inhabit a certain place." "Authors" (Choice *B*) are a subset of people who might be inhabiting a place; "cultures" (Choice *C*) encompass more than just the people themselves living somewhere; and "persons" (Choice *D*) applies to a specific, small number of people.

42. B: Choice *B* is the correct answer. Style and tone are important aspects of essay writing, and writing that distracts the reader from the point being made is to be avoided. Readers know when a writer is trying too hard, as is the case here.

43. D: This sentence has three pieces of information: there are *two levels of significance* to the *concept of three-ness* that is *applied by the authors*. The reader already knows that the authors of the sagas are the people who are applying the concept of three, so that piece of information does not need to be repeated. Choice *D* is the only choice that omits mentioning the authors and has no flaws. Choice *A* erroneously replaces "in" with "to," which does not make the sentence any clearer. Choice *B* indicates that the application, not the significance, has two levels. Choice *C* is close to the correct answer, but it would be simpler if it did not include the authors.

44. C: If the new paragraph appeared before the final paragraph, the order of topics would flow logically:

- Introduction of concept of three
- First level of significance
- Second level of significance
- Dual purpose of three-ness
- How the application of three-ness evolved over time

45. A: Choice *A* is the clearest rewrite. It replaces the "not only...but also..." structure with two sentences that are relatively simple to comprehend. Choice *B*, in the interest of brevity, leaves out two concepts: tempo and balance. In Choice *C*, the conjunction "but," which means "on the contrary," throws the intended meaning of the passage into question. Choice *D* is written in the passive voice (the subject is the recipient of the action), which is less direct than the active voice.

Math

1. D: The volume for a cylinder is found by using the formula:

$$V = \pi r^2 h = \pi(2 \text{ in})^2 \times 3.5 \text{ in} = 43.96 \text{ in}^3$$

2. D: The argument of a logarithmic function has to be greater than or equal to zero. Basically, one cannot take the logarithm of a negative number or 0. Therefore, to find the domain, set the argument greater than 0 and solve the inequality. This results in $x - 2 > 0$, or $x > 2$. Therefore, in order to obtain an output of the function, the number plugged into the function must be greater than 2. This domain is represented as $(2, \infty)$.

3. B: The equivalent equation form of the function is $y = 10^{x+3}$, with base 10. The expression can be written in logarithmic form as $\log_{10} y = x + 3$. A logarithm with base 10 is known as the common logarithm and is written without the base. Adding -3 to both sides results in,

$$-3 + \log y = x$$

4. A: Surface area is a type of area, which means it is measured in square units. Cubic units are used to describe volume, which has three dimensions multiplied by one another. Quartic units describe measurements multiplied in four dimensions.

5. C: Plugging the function into the formula results in:

$$\frac{4(x + h) - 2 - (4x - 2)}{h}$$

which is simplified to:

$$\frac{4x + 4h - 2 - 4x + 2}{h} = \frac{4h}{h} = 4$$

This value is also equal to the derivative of the given function. The derivative of a linear function is its slope.

6. C: By switching from a radical expression to rational exponents:

$$\sqrt[4]{x^6} = x^{\frac{6}{4}} = x^{\frac{3}{2}}$$

Also, properties of exponents can be used to simplify $\frac{x}{x^3}$ into:

$$x^{1-3} = x^{-2} = \frac{1}{x^2}$$

The other terms can be left alone, resulting in an equivalent expression:

$$x^{\frac{3}{2}} - \frac{1}{x^2} + x - 2$$

7. B: The expression under the radical can be factored into perfect squares as such:

$$\sqrt{100 \times 2 \times x^6 \times y^6 \times y \times z^2}$$

Using rational exponents, this is the same as:

$$(100 \times 2 \times x^6 \times y^6 \times y \times z^2)^{\frac{1}{2}}$$

$$100^{\frac{1}{2}} \times 2^{\frac{1}{2}} \times x^{\frac{6}{2}} \times y^{\frac{6}{2}} \times y^{\frac{1}{2}} \times z^{\frac{2}{2}}$$

$$\sqrt{100} \times \sqrt{2} \times x^3 \times \sqrt{y} \times z$$

$$10x^3y^3z\sqrt{2y}$$

8. C: First, the slope of the line must be found. This is equal to the change in y over the change in x, given the two points. Therefore, the slope is -6. The slope and one of the points are then plugged into the point-slope form of a line:

$$y - y_1 = m(x - x_1)$$

This results in:

$$y - 7 = -6(x + 3)$$

The -6 is distributed and the equation is solved for y to obtain $y = -6x - 11$.

9. A: A vertical line has the same x value for any point on the line. Other points on the line would be (1, 3), (1, 5), (1, 9), etc. Mathematically, this is written as $x = 1$. A vertical line is always of the form $x = a$ for some constant a.

10. B: To rationalize the denominator, the conjugate of the denominator must be multiplied by both the numerator and the denominator of the expression. This will result in a denominator without radicals. The conjugate of the denominator is $x + \sqrt{7}$. Multiplying this by both the numerator and denominator results in:

$$\frac{49(x + \sqrt{7})}{(x - \sqrt{7})(x + \sqrt{7})} = \frac{49(x + \sqrt{7})}{x^2 - 7}$$

11. B: In order to find $h(-5x)$, plug $-5x$ in for x in the given function. Therefore,

$$h(-5x) = -5(-5x) - 7 = 25x - 7$$

12. A: Plugging 2 in for x in the function results in:

$$g(2) = a(2^2) + 4(2) + 9 = 4a + 17$$

Because this expression is equal to -7, it is true that $4a + 17 = -7$. Therefore, $4a = -24$ and $a = -6$.

13. D: In order to add the two rational expressions, a common denominator must be used. The common denominator is $(x+5)(x-5)$. Multiply by fractions equal to 1 to build the common denominator. Multiply the first expression times $\frac{x-5}{x-5}$ and the second times $\frac{x+5}{x+5}$, and then subtract to obtain:

$$\frac{(4x-2)(x-5)}{(x+5)(x-5)} - \frac{(2x-1)(x+5)}{(x-5)(x+5)}$$

$$\frac{4x^2-20x-2x+10}{(x+5)(x-5)} - \frac{2x^2+10x-x-5}{(x+5)(x-5)}$$

$$\frac{2x^2-31x+15}{x^2-25}$$

14. A: If $M = 5x^2 - 4x + 2$ and $N = -2x^2 - 6x + 1$, then:

$$3M - \frac{3}{4}N = 3(5x^2 - 4x + 2) - \frac{3}{4}(-2x^2 - 6x + 1)$$

Set the coefficients to a common denominator:

$$3M - \frac{3}{4}N = \frac{12}{4}(5x^2 - 4x + 2) - \frac{3}{4}(-2x^2 - 6x + 1)$$

Multiply each term of the polynomials by their coefficients and simplify:

$$3M - \frac{3}{4}N = \frac{60x^2}{4} - \frac{48x}{4} + \frac{24}{4} + \frac{6x^2}{4} + \frac{18x}{4} - \frac{3}{4}$$

$$3M - \frac{3}{4}N = \frac{66x^2}{4} - \frac{30x}{4} + \frac{21}{4}$$

$$3M - \frac{3}{4}N = \frac{33x^2}{2} - \frac{15x}{2} + \frac{21}{4}$$

15. C: Solving the equation for B results in:

$$B = \frac{81x^2}{9x+2} - \frac{4}{9x+2}$$

$$B = \frac{81x^2-4}{9x+2}$$

$$B = \frac{(9x+2)(9x-2)}{9x+2}$$

$$B = 9x - 2$$

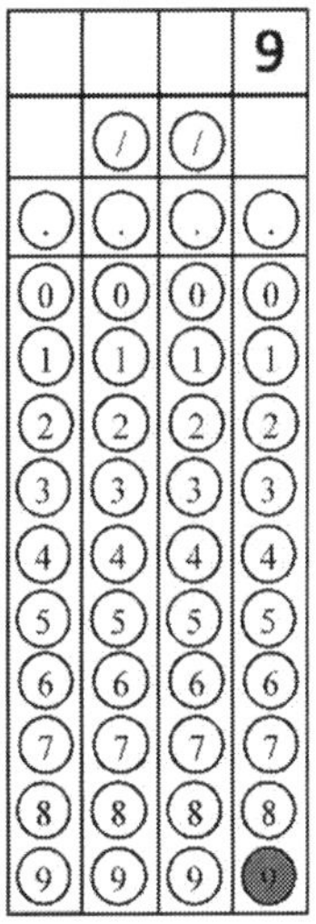

16. The area of the shaded region is calculated in a few steps. First, the area of the rectangle is found using the formula:

$$A = length \times width = 6 \text{ m} \times 2 \text{ m} = 12 \text{ m}^2$$

Second, the area of the triangle is found using the formula:

$$A = \frac{1}{2} \times base \times height = \frac{1}{2} \times 3 \text{ m} \times 2 \text{ m} = 3 \text{ m}^2$$

The last step is to take the rectangle area and subtract the triangle area. The area of the shaded region is:

$$A = 12 \text{ m}^2 - 3 \text{ m}^2 = 9 \text{ m}^2$$

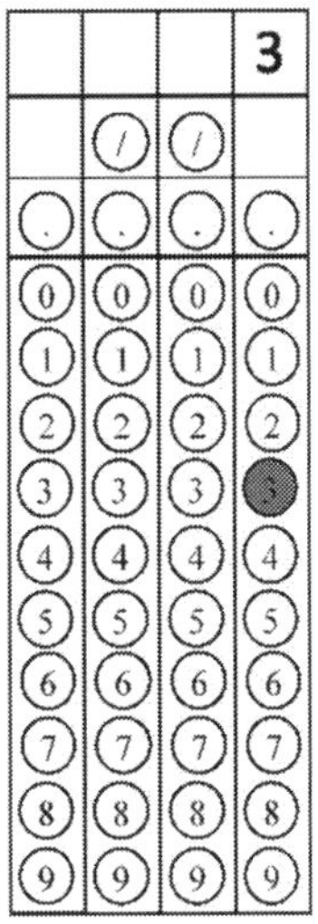

17. One way to determine the number of zeros is to count the number of sign changes in coefficients in the polynomial. This results in the number of possible positive zeros. The coefficients are 1, -3, 2, 1, and -3, so the sign changes from 1 to -3, -3 to 2, and 1 to -3, a total of 3 times. Therefore, there are at most 3 positive zeros.

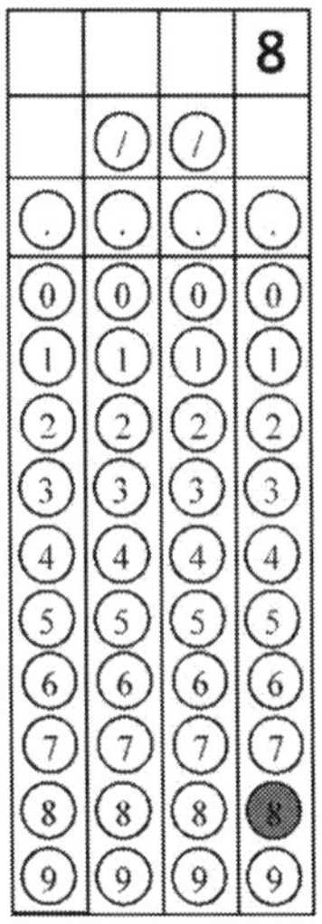

18. The first step is to simplify the expression. The second term, $16^{\frac{1}{2}}$, can be rewritten as $\sqrt{16}$, since fractional exponents represent roots. This can then be simplified to 4, because 16 is the perfect square of 4. The first term can be evaluated using the power rule for exponents: an exponent to a second exponent is multiplied by that second exponent. The first term can thus be rewritten as:

$$16^{\frac{1}{4}} = \left(16^{\frac{1}{2}}\right)^{\frac{1}{2}} = 4^{\frac{1}{2}} = 2$$

Combining these terms, the expression is evaluated as follows:

$$16^{\frac{1}{4}} \times 16^{\frac{1}{2}} = 2 \times 4 = 8$$

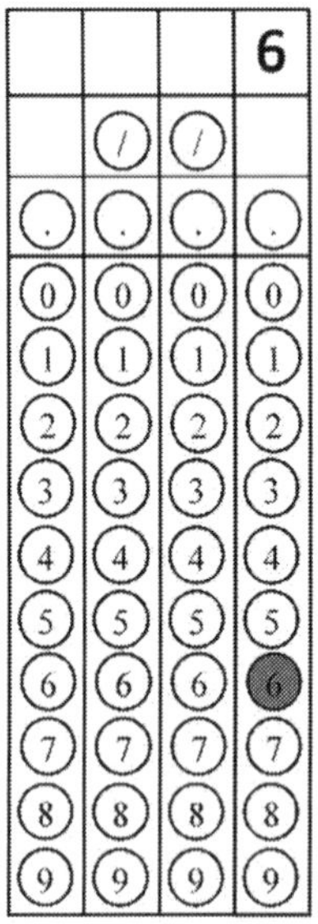

19. 6: The function is undefined where the denominator is equal to 0. Setting the denominator equal to 0 results in

$$(x-9)^2 + 6(x-9) + 9 = 0$$

This is equivalent to the square of a binomial $(a+b)^2 = a^2 + 2ab + b^2$ for $a = x - 9$ and $b = 3$. It factors into:

$$\left((x-9)+3\right)^2 = (x-6)^2 = 0$$

which has solution $x = 6$. Therefore, the function is undefined at 6.

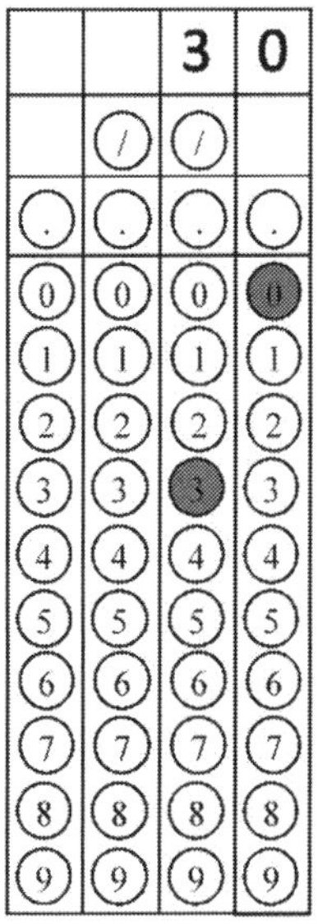

20. 30: From the table, we can see that $f(6) = 2$ and $g(6) = -8$. Therefore:

$$h(6) = -4(-8) - 2$$

$$h(6) = 32 - 2$$

$$h(6) = 30$$

21. D: This is a one-step real-world application problem. The unknown quantity is the number of cases of cola to be purchased. Let x be equal to this amount. Because each case costs $3.50, the total number of cases multiplied by $3.50 must equal $40. This translates to the mathematical equation $3.5x = 40$. Divide both sides by 3.5 to obtain $x = 11.4286$, which has been rounded to four decimal places. Because cases are sold whole, and there is not enough money to purchase 12 cases, 11 cases is the correct answer.

22. A: Rounding can be used to find the best approximation. All of the values can be rounded to the nearest thousand. 15,412 SUVs can be rounded to 15,000. 25,815 station wagons can be rounded to 26,000. 50,412 sedans can be rounded to 50,000. 8,123 trucks can be rounded to 8,000. Finally, 18,312 hybrids can be rounded to 18,000. The sum of the rounded values is 117,000, which is closest to 120,000.

23. D: There are 52 weeks in a year, and if the family spends $105 each week, that amount is close to $100. A good approximation is $100 a week for 50 weeks, which is found through the product:

$$50 \times 100 = \$5{,}000$$

24. D: There were 48 total bags of apples sold. If 9 bags were Granny Smith and the rest were Red Delicious, then $48 - 9 = 39$ bags were Red Delicious. Therefore, the ratio of Granny Smith to Red Delicious is 9:39.

25. B: The average rate of change is found by calculating the difference in dollars over the elapsed time. Therefore, the rate of change is equal to $(\$4{,}900 - \$4{,}000) \div 3$ months, which is equal to $\$900 \div 3$, or $300 per month.

26. A: Let x be the unknown, the number of hours Erin can work. We know Katie works $2x$, and the sum of all hours is less than 21. Therefore, $x + 2x < 21$, which simplifies into $3x < 21$. Solving this results in the inequality $x < 7$ after dividing both sides by 3. Therefore, Erin worked less than 7 hours.

27. D: The area for a rectangle is found by multiplying the length by the width. The area is also measured in square units, so the correct answer is Choice *D*. The answer of 26 is the perimeter. The answer of 13 is found by adding the two dimensions instead of multiplying.

28. A: The surface area for a cylinder is the sum of the two circle bases and the rectangle formed on the side. This is easily seen in the net of a cylinder.

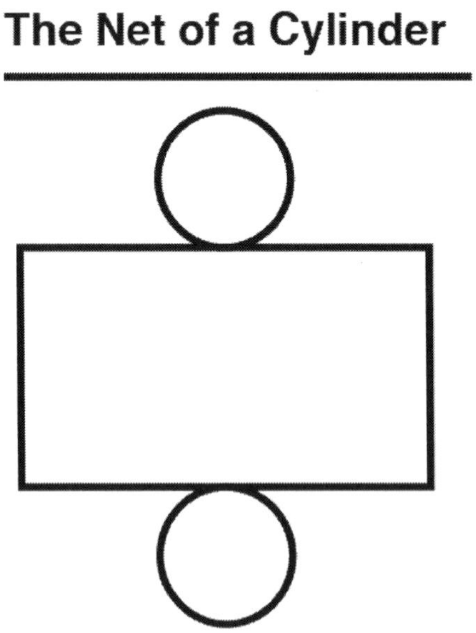

The area of a circle is found by multiplying π times the radius squared. The rectangle's area is found by multiplying the circumference of the circle by the height. The equation $SA = 2\pi \times 5 \times 10 + 2(\pi 5^2)$ shows the area of the rectangle as $2\pi \times 5 \times 10$, which yields 314. The area of the bases is found by $\pi 5^2$, which yields 78.5, then multiplied by 2 for the two bases.

29. A: A proportion should be used to solve this problem. The ratio of tagged to total deer in each instance is set equal, and the unknown quantity is a variable x. The proportion is:

$$\frac{300}{x} = \frac{5}{400}$$

Cross-multiplying gives $120{,}000 = 5x$, and dividing through by 5 results in 24,000.

30. B: First, the common factor 2 can be factored out of both terms, resulting in:

$$2(y^3 - 64)$$

The resulting binomial is a difference of cubes that can be factored using the rule:

$$a^3 - b^3 = (a - b)(a^2 + ab + b^2)$$

with $a = y$ and $b = 4$. Therefore, the result is:

$$2(y - 4)(y^2 + 4y + 16)$$

31. D: The exponential rules $(ab)^m = a^m b^m$ and $(a^m)^n = a^{mn}$ can be used to rewrite the expression as:

$$4^4 y^{12} \times 3^2 y^{14}$$

The coefficients are multiplied together and the exponential rule $a^m a^n = a^{m+n}$ is then used to obtain the simplified form $2{,}304y^{26}$.

32. B: The number of representatives varies directly with the population, so the equation necessary is $N = k \times P$, where N is number of representatives, k is the variation constant, and P is total population in millions. Plugging in the information for New York allows k to be solved for.

This process gives $27 = k \times 19.8$, so $k = 1.36$. Therefore, the formula for number of representatives given total population in millions is $N = 1.36 \times P$. Plugging in $P = 11.8$ for Ohio results in $N = 16.05$, which rounds to 16 total representatives.

33. B: The outlier is 35. When a small outlier is removed from a data set, the mean and the median increase. The first step in this process is to identify the outlier, which is the number that lies away from the given set. Once the outlier is identified, the mean and median can be recalculated. The mean will be affected because it averages all of the numbers. The median will be affected because it finds the middle number, which is subject to change because a number is lost. The mode will most likely not change because it is the number that occurs the most, which will not be the outlier if there is only one outlier.

34. C: The mean is found by adding all the times together and dividing by the number of times recorded.

$$25 + 18 + 23 + 28 + 30 + 22.5 + 23 + 33 + 20 = 222.5$$

$$\frac{222.5}{9} = 24.722$$

Rounding to the nearest minute, the mean is 25 minutes.

35. C: The mode is the time from the data set that occurs most often. The number 23 occurs twice in the data set, while all others occur only once, so the mode is 23 minutes.

36. A: To find the median of a data set, you must first list the numbers from smallest to largest, and then find the number in the middle. If there are two numbers in the middle, add the two numbers in the middle together and divide by 2. Putting this list in order from smallest to greatest yields 18, 20, 22.5, 23, 23, 25, 28, 30, and 33, where 23 is the middle number, so 23 minutes is the median.

37. B: Look on the horizontal axis to find 3:00 p.m. Move up from 3:00 p.m. to reach the dot on the graph. Move horizontally to the left to the horizontal axis to between 20 and 25; the best answer choice is 22. The answer of 25 is too high above the projected time on the graph, and the answers of 20 and 16 degrees are too low.

38. B: The variable x represents the number of years after 1980. The year 2011 was 31 years after 1980, so plug 31 into the function to obtain:

$$f(31) = 3.1 \times 31 + 240 = 336.1$$

This value rounds to 336 and represents 336 million people.

39. A: The zeros of a polynomial function are the x-values where the graph crosses the x-axis, or where $y = 0$. Therefore, set $y = 0$ and solve the polynomial equation. This quadratic can be solved using factoring, as follows:

$$0 = 2x^2 - 12x + 16$$

$$0 = 2(x^2 - 6x + 8) = 2(x - 4)(x - 2)$$

Setting both factors equal to zero results in the two solutions $x = 4$ and $x = 2$, which are the zeros of the original function.

40. A: In order to solve for x, first clear the equation of fractions, by multiplying each side times $\frac{4}{3}$. This step results in the equation:

$$x - \frac{11}{15} = 15$$

Then, add $\frac{11}{15}$ to both sides, which results in:

$$x = 15 + \frac{11}{15} = \frac{225 + 11}{15} = \frac{236}{15}$$

The equivalent decimal form, rounded to the nearest hundredth, is 15.73.

41. C: Let x be the cost of the pencil. Therefore, $3x - 0.50$ is the cost of the pen. Together, they sum up to $5.10, which is:

$$x + (3x - 0.50) = 5.10$$

Solve for x to find the cost of the pencil:

$$x + 3x = 5.10 + 0.50$$

$$4x = 5.60$$

$$x = 1.40$$

Therefore, the pencil costs $1.40 and the pen costs $3(1.40) - 0.50 = 3.70$, or $3.70. Thus, the pen costs $\$3.70 - \$1.40 = \$2.30$ more than the pen.

42. B: The slope of the line given is:

$$m = \frac{6.75 - 3.95}{7.12 - 5.27} = \frac{2.8}{1.4} = 2$$

Plugging this and the first point into the point-slope formula for the equation of a line results in:

$$y - 3.95 = 2(x - 5.72)$$

which has equivalent form:

$$y = 2x - 7.49$$

43. A: Solving the first equation for x results in the equation $x = 6 - y$. This can be plugged in to x in the second equation to obtain:

$$5(6 - y) - 6y = 8$$

Distributing and collecting like terms gives an equivalent equation $30 - 11y = 8$. Subtracting 30 from both sides and dividing by -11 results in the solution $y = 2$. Therefore, $x = 6 - 2 = 4$. Finally, plugging these values into the original fraction yields:

$$\frac{x}{y} = \frac{4}{2} = 2$$

44. C: Because the line goes through the origin, the slope is equal to

$$m = \frac{-50 - 0}{a - 0} = -\frac{50}{a}$$

and

$$m = \frac{a - 0}{-2 - 0} = -\frac{a}{2}$$

Because slope is a single value, these two values are equal. Therefore:

$$-\frac{50}{a} = -\frac{a}{2}$$

Cross-multiply to isolate the variable to one side and solve for a:

$$-50 \times 2 = -a^2$$

$$a^2 = 100$$

$$a = 10$$

Therefore, the value of a is 10.

45. C: To determine the solution, solve the inequality for c. Subtracting $8.5c$ from both sides, results in

$$-2c - 11.2 \geq 9$$

Then, add 11.2 to both sides and divide by -2, making sure to flip the sign of the inequality due to the division by a negative number. The solution is $c \leq -10.1$. The only value that does not satisfy this inequality is -10.

46. B: If x represents the online students enrolled from 2010 to 2015, then 15 times this amount is equal to $15x$. Because there were 2400 students from 2015 to 2020, $15x = 2400$. Dividing by 15 results in $x = 160$. Therefore, there were 160 online students enrolled from 2010 to 2015.

47. A: Rewriting the system in standard form results in

$$48x + cy = -25$$

$$12dx - 10y = -11$$

Because the system of equation has no solution, the x —coefficients must be equal and the y —coefficients must be equal. Therefore, $12d = 48$, so $d = 4$. Also, $c = -10$. Therefore,

$$|c - d| = |-10 - 4| = |-14| = 14$$

48. C: Because each multiple-choice question is worth 2.5 points, and she answers 15 of them correctly, her total score from multiple-choice questions is $2.5(15) = 37.5$. Because each short answer question is worth 4 points, and she answers x of them correctly, her total score from short answer questions is $4x$. Adding these together results in $4x + 37.5$, her total score.

49. C: Multiplying the first expression results in:

$$(17.64x^2 - 26.88x + 10.24) - (0.54x^2 - 1.21x + 11.81) = 17.1x^2 - 25.67x - 1.57$$

50. B: Squaring both sides results in:

$$4x^2 - 12 = 1012$$

Add 12 to both sides and divide by 4 to obtain $x^2 = 256$. This equation has two solutions: $x = \pm 16$. However, because $x > 0$, the correct answer is 16.

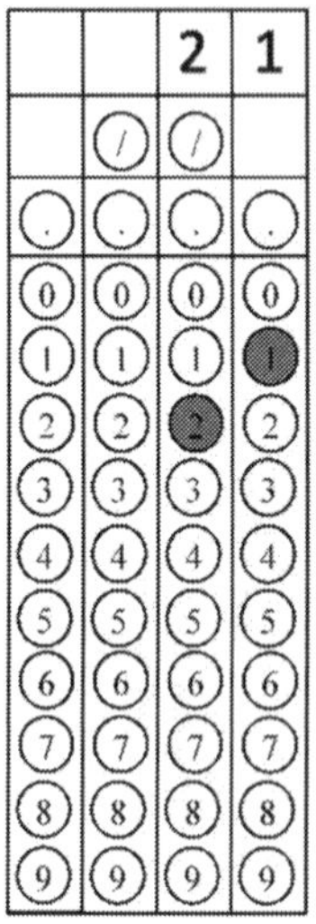

51. Gina answered 60% of 35 questions correctly; 60% can be expressed as the decimal 0.60. Therefore, she answered $0.60 \times 35 = 21$ questions correctly.

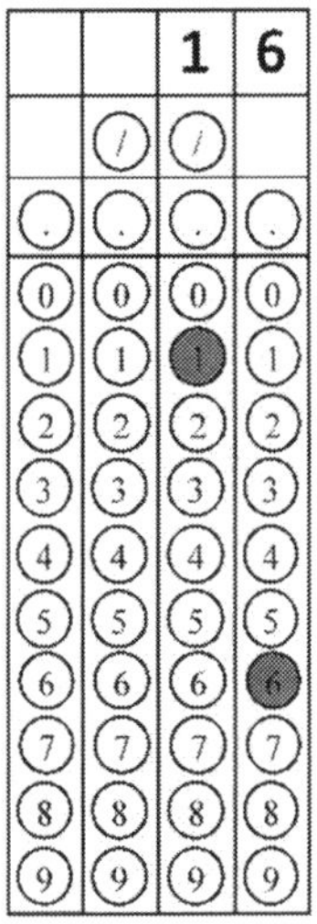

52. The unknown quantity is the number of total questions on the test. Let x be equal to this unknown quantity. Therefore, $0.75x = 12$. Divide both sides by 0.75 to obtain $x = 16$.

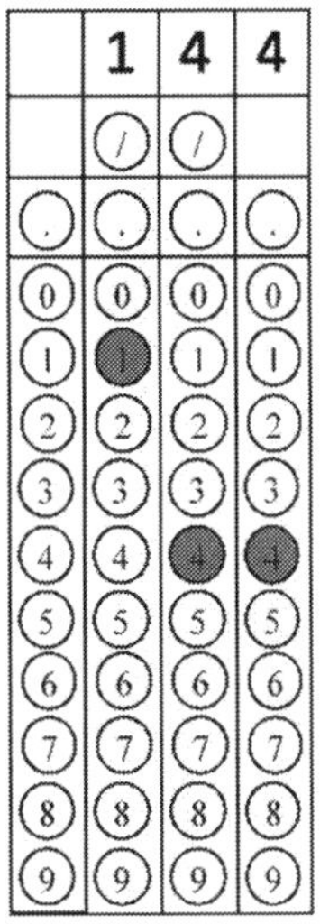

53. This problem involves ratios and percentages. 12 packets needed for every 5 people is equivalent to the ratio $\frac{12}{5}$. The unknown amount x is the number of ketchup packets needed for 60 people. The proportion $\frac{12}{5} = \frac{x}{60}$ must be solved. Cross-multiply to obtain $12 \times 60 = 5x$. Therefore, $720 = 5x$. Divide each side by 5 to obtain $x = 144$.

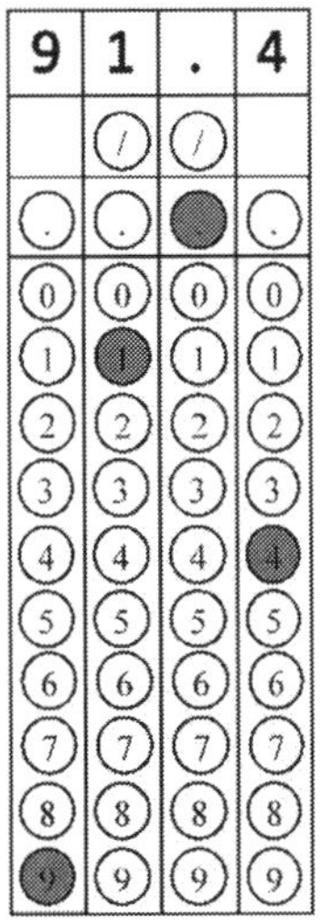

54. The conversion between feet and centimeters requires a middle term. As there are 2.54 centimeters in 1 inch, the conversion between inches and feet must be found. As there are 12 inches in a foot, the fractions can be set up as follows:

$$3 \text{ ft} \times \frac{12 \text{ in}}{1 \text{ ft}} \times \frac{2.54 \text{ cm}}{1 \text{ in}}$$

The feet and inches cancel out to leave only centimeters for the answer. The numbers are calculated across the top and bottom to yield:

$$\frac{3 \times 12 \times 2.54}{1 \times 1} = 91.44$$

The number and units used together form the answer of 91.4 cm.

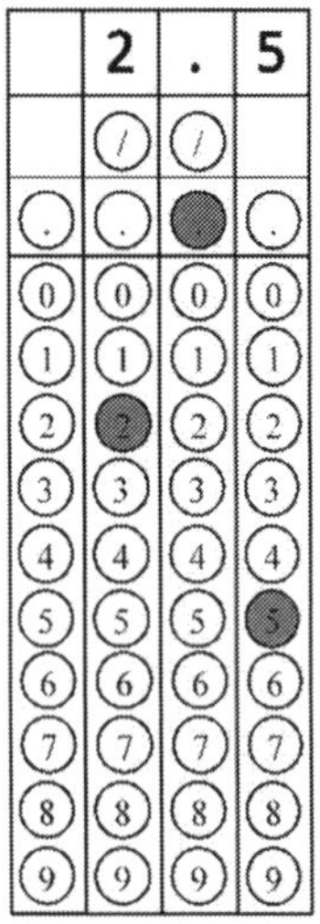

55. Let x be the unknown number. The difference indicates subtraction, and sum represents addition. To triple the difference, it is multiplied by 3. The problem can be expressed as the following equation:

$$3(5 - x) = x + 5$$

Distributing the 3 results in:

$$15 - 3x = x + 5$$

Subtract 5 from both sides, add $3x$ to both sides, and then divide both sides by 4. This results in:

$$x = \frac{10}{4} = \frac{5}{2} = 2.5$$

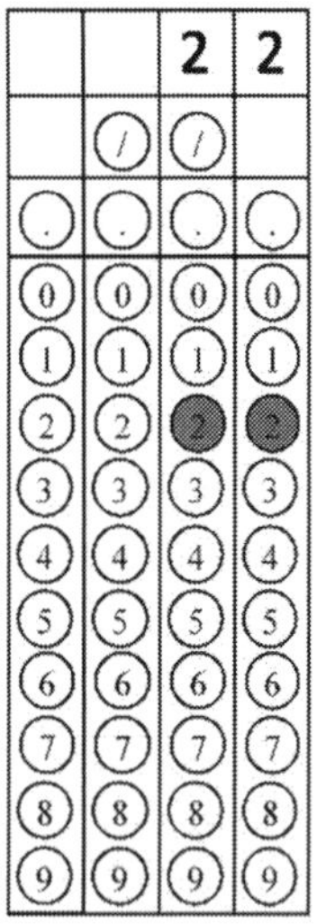

56. The perimeter is found by adding the length of all the exterior sides. When the given dimensions are added, the perimeter is 22 meters. The equation to find the perimeter can be:

$$P = 5 + 1.5 + 1.2 + 4.5 + 3.8 + 6 = 22$$

The last two dimensions can be found by subtracting 1.2 from 5, and adding 1.5 and 4.5, respectively.

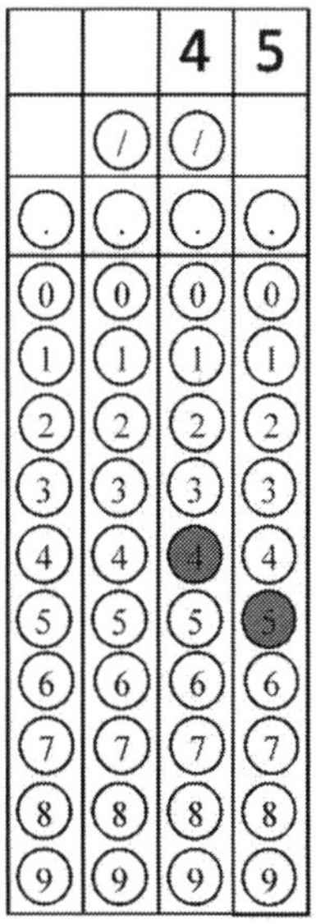

57. 45: To find the y −intercept, plug 0 in for x and solve for y. Therefore,

$$\frac{4}{3}(0) - \frac{11}{9}y = -55$$

$$-\frac{11}{9}y = -55$$

Multiplying both sides by $-\frac{9}{11}$ gives:

$$y = \frac{495}{11} = 45$$

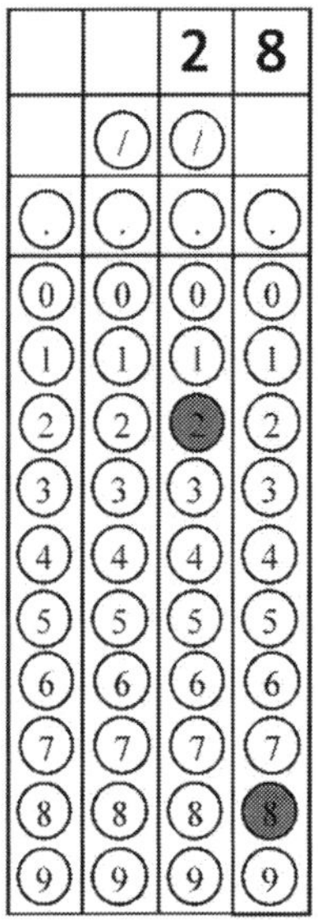

58. 28: In order to answer this question, manipulate the given equation to look like the terms in the expression. Therefore, subtract $0.25b$ from both sides and reorder the terms to obtain:

$$0.25b + 8.75x = 7$$

Then, divide both sides by 0.25. The final result is:

$$b + 35x = 28$$

Greetings!

First, we would like to give a huge "thank you" for choosing us and this study guide for your SAT exam. We hope that it will lead you to success on this exam and for your years to come.

Our team has tried to make your preparations as thorough as possible by covering all of the topics you should be expected to know. In addition, our writers attempted to create practice questions identical to what you will see on the day of your actual test. We have also included many test-taking strategies to help you learn the material, maintain the knowledge, and take the test with confidence.

We strive for excellence in our products, and if you have any comments or concerns over the quality of something in this study guide, please send us an email so that we may improve.

As you continue forward in life, we would like to remain alongside you with other books and study guides in our library, such as;

ACCUPLACER:

amazon.com/dp/1628457724

We are continually producing and updating study guides in several different subjects. If you are looking for something in particular, all of our products are available on Amazon. You may also send us an email!

Sincerely,
APEX Test Prep
info@apexprep.com

Printed in Great Britain
by Amazon

66122816R00156